Freshwater studies

PRACTICAL ECOLOGY SERIES

Seashore Studies
Urban Ecology
Grassland Studies
Freshwater Studies
Woodland Studies
Upland and Moorland Studies
Salt Marsh and Estuary Studies

Series editor

Morton Jenkins, B.Sc., M.I.Biol.

Head of Science, Howardian High School, Cardiff
Chief examiner W.J.E.C. in CSE Biology
Examiner C.U.E.S.

Freshwater studies

John H. R. Gee, B.Sc., D.Phil.

Department of Zoology, The University College of Wales, Aberystwyth

London
GEORGE ALLEN & UNWIN

Boston   Sydney

George Allen & Unwin (Publishers) Ltd,
40 Museum Street, London WC1A 1LU, UK

George Allen & Unwin (Publishers) Ltd,
Park Lane, Hemel Hempstead, Herts HP2 4TE, UK

Allen & Unwin Inc.,
Fifty Cross Street, Winchester, Mass. 01890, USA

George Allen & Unwin Australia Pty Ltd,
8 Napier Street, North Sydney, NSW 2060, Australia

First published in 1986

ISSN 0261–0493

British Library Cataloguing in Publication Data

Gee, John H.R.
    Freshwater studies.—(Practical ecology
    series, ISSN 0261-0493; 4)
    1. Freshwater ecology—Juvenile literature
    I. Title   II. Series
    574.5′2632   QH541.5.F7

    ISBN 0-04-574024-0

Library of Congress Cataloging in Publication Data

Gee, John H. R.
    Freshwater studies.
    (Practical ecology series)
    Bibliography: p.
    Includes index.
    1. Freshwater ecology—Problems, exercises, etc.
2. Freshwater ecology—Great Britain—Problems,
exercises, etc. 3. Freshwater ecology—Field work.
I. Title. II. Series: Practical ecology.
QH541.5.F7G44 1985   574.5′2632   85-15668
ISBN 0-04-574024-0   (U.S.: alk. paper)

Set in 10 on 12 point Times by
Mathematical Composition Setters Ltd, Salisbury, UK
and printed in Great Britain by
Mackays of Chatham

Foreword

The aim of this series is to provide students and teachers of Advanced level biological science with ideas for a practical approach to ecology. Each book deals with a particular ecosystem and has been written by an experienced teacher who has had a particular interest in organising and teaching field work. The texts include:

(a) an introduction to the ecosystem studied;
(b) keys necessary for the identification of organisms used in the practical work;
(c) background information relevant to field and laboratory studies;
(d) descriptions of methods and techniques used in the practical exercises;
(e) lists of materials needed for the practical work described;
(f) realistic suggestions for the amount of time necessary to complete each exercise;
(g) a series of questions to be answered with knowledge gained from an investigatory approach to the study;
(h) a bibliography for further reference.

Throughout the series emphasis is placed on *understanding* the ecology, rather than on compiling lists, of organisms. The identification of types, with the use of keys, is intended to be a means to an end rather than an end in itself.

Morton Jenkins
Series editor

Preface

There is a freshwater habitat within easy reach of every school and college in Britain. Streams or ditches can often be found within school grounds and many establishments have created their own garden ponds. Freshwater environments provide scope for investigating almost every aspect of ecology, so they make ideal sites for the ecological fieldwork that is part of most Advanced level biology syllabuses.

I have tried to give this book a strong ecological framework and to cover as many ecological concepts in the exercises as possible. None of the exercises requires elaborate or expensive apparatus and the wide range of time requirements can be used by teachers to select an exercise to fit the time available.

Tackling practical work in an unfamiliar environment can be daunting; I hope to persuade the reader that the effort is worth while and to provide the information and advice necessary for a successful practical investigation of the ecology of fresh water.

John Gee

Acknowledgements

I am grateful to E. Schweizerbart'sche Verlagsbuchhandlung, and to Dr J. M. Elliott and the Lake District Special Planning Board for permission to reproduce Figures 6 and 25, respectively. The cover photograph of the Afon Rheidol is the work of R. A. Moore and D. Williams assisted with electron microscopy. Three reviewers made suggestions for improvements to the text; Dr Alan Hildrew was particularly helpful in this respect. Finally, my wife, Anne, tolerated the intrusion into family life admirably and provided the necessary encouragement when lethargy and inertia threatened the writing process.

Contents

Introduction

Water is in continuous circulation between the sea, the atmosphere and the land. Evaporation from the sea provides the moisture in the atmosphere which falls as rain or snow when oceanic air meets a land mass. Drainage from the land mass eventually returns to the sea through the rivers. This is called the hydrological cycle.

Unpolluted rain contains very small amounts of salts, having originally been distilled from the surface of the sea. As it drains through the landscape, rain water dissolves and erodes the rocks and soils. The amounts of mineral salts that dissolve in the water are dependent on the geology of the land. Dissolved salts are carried to the sea and concentrated by evaporation. Sea water contains about $35\,\mathrm{g\,l^{-1}}$ of dissolved salts, a figure that varies little across the globe, but the concentration of salts in fresh water is always much lower and can vary over several orders of magnitude. For the purposes of definition, water containing less than $5\,\mathrm{g\,l^{-1}}$ is regarded as fresh, although this figure is somewhat arbitrary.

Rain falling on soil percolates downwards to the level of the water table, the level below which all the spaces between the soil particles are filled with water. In places where the surface of the soil is below the local water table, water appears on the surface of the ground to form streams. Streams join to form larger streams and rivers in a branching drainage network. When rivers approach the sea and are influenced by tidal movements they become estuaries.

A pond or lake forms on a river when an obstruction blocks the progress of the water down hill. Water backs up behind the obstruction until it can flow over or around it, so lakes are simply places where the water is temporarily delayed. The distinction between a pond and a lake is arbitrarily based on size, and to some extent the distinction between lakes and rivers is also artificial. Water passes through a river channel relatively quickly but passes much more slowly through the basin of a lake. The difference is only a matter of degree; the biotic and abiotic conditions 'typical' of a river can often be found in a lake, and vice versa. Since lake basins are continually being filled in by the sediment carried to them by rivers, they eventually disappear completely. On a geological timescale lakes are relatively ephemeral features on the landscape.

A characteristic set of organisms has evolved to exploit the ecological possibilities offered by freshwater environments. Whereas the shallow seas are dominated by the large algae or seaweeds, the large plants in fresh water are mainly drawn from the Spermatophyta, or flowering plants. Although marine and freshwater habitats share many groups of microscopic plants, the species are different, as is the relative contribution made by each group.

Animal communities in fresh water are often dominated by the insects and their larvae. Insects are basically terrestrial animals that have evolved to fill some of the niches in fresh water that are occupied by Crustacea in the sea. Many groups of animals that are important in the sea are absent from fresh water, presumably having failed to meet the exacting requirements of freshwater life. In this category fall the echinoderms, tunicates, cephalopod molluscs and polychaete worms. On the other hand, some common freshwater animals, such as leeches and oligochaete worms, are less important or virtually absent from the sea.

Of all aquatic environments, fresh water is probably the one most affected by man; indeed many freshwater habitats in Britain are man-made. These include the power generation dams of highland areas, the reservoirs and gravel pits of the lowland landscape and the historic peat diggings that form the Norfolk Broads. Even among natural water bodies there are few that are free of human influences. Many have been modified in the interests of fishing, navigation or flood control, and incidences of industrial or domestic pollution are distressingly frequent.

Freshwater environments encompass a wide range of physical and chemical conditions, both natural and man-made. They contain plants and animals that are highly adapted to the rigours of freshwater life, but communities that are not so rich in species as to defy ecological analysis. In Britain we are fortunate to be heirs to a long tradition of study in freshwater biology. Many of the problems of identification that often precede an ecological investigation have already been solved. For these reasons, fresh water offers ecologists exciting opportunities to study the functioning of a natural environment.

Keys

Progress in science depends to a large extent on the exchange of information between scientists. A single biologist could conceivably investigate an organism which he could not name, but he would find it difficult to compare his results with those of anyone else. The ability to give an internationally recognised name to an organism is therefore necessary for useful work on its biology.

A set of keys which enable identification of all the British freshwater organisms to the species level is far beyond the scope of this book. Indeed, for some groups adequate species keys do not exist. Instead the keys in this chapter are designed for identification of common British freshwater invertebrates to the level required by the practical exercises. In some cases a species name is given, whilst in others the end point is a broader taxonomic group. A name in **bold** type indicates an end point. Colloquial English names are given in brackets. Generally, identification is taken to the point at which major distinctions, in terms of ecological roles, can be made between the groups.

As far as possible the keys have been designed so that the diagnostic features can be seen by eye or with the aid of a hand lens. If a low-power binocular dissecting microscope is available, so much the better. Often the interested student will be able to take identification further by consulting the sources listed in the Bibliography.

Key 1: a preliminary key to the common British freshwater invertebrates

1 Animal less than 2 mm long and *either* unicellular *or* composed of a group of similar cells. **Protozoa**
   Clearly composed of many cells with different shapes and functions.
   Metazoa **2**

2 Texture spongy with many small holes in outer surface. Occurs as immobile green or grey growth on stones.
   Phylum **Porifera** (freshwater sponge) Family **Spongillidae**
   Individuals are of similar, well defined shape, usually symmetrical. Mobile or capable of rapid contraction. **3**

3 Soft, green or brown tubular body surmounted by a ring of tentacles. Sessile, but contracts when disturbed.
   Phylum **Cnidaria**, genus *Hydra*
   Without a ring of contractile tentacles. **4**

4 Less than 3 mm long, with a crown of beating cilia. May be attached by a two-toed 'foot'. Sometimes has a thickened or ornamented cuticle. (Rotifer) Phylum **Rotifera**

Without a crown of cilia. Usually longer than 3 mm. 5

5 Thin, worm-like animal. No obvious hard parts. If segmented, has more than 14 segments. 6

Either not segmented *or* having less than 14 segments *or* some part of the body chitinised and hard. 12

6 Segmented. Phylum Annelida 7

Not segmented. 11

7 Sucker on ventral side at each end. Moves by 'looping' or occasionally by swimming. (leeches) Class Hirudinea 8

No suckers. Movement worm-like. (worms) Class **Oligochaeta**

8 *Either* more than eight eyes *or* with eight eyes arranged in two parallel, longitudinal rows. 9

Either more than eight eyes *or* with eight eyes not arranged as above. Families **Hirudinidae** and **Erpobdellidae**

9 Head and anterior sucker wider than 'neck'. Genera *Piscicola* and *Hemiclepsis*

Head and anterior sucker not clearly wider than 'neck'. 10

10 Two eyes, small leathery plate on dorsal surface near head. *Helobdella stagnalis*

Six eyes. *Glossiphonia* **spp.**

Eight eyes. *Theromyzon tessulatum*

11 Dorsoventrally flattened. Blunt ended. Moves by gliding on a mucous trail. May have eyes. (flatworms) Phylum **Platyhelminthes**, Class **Turbellaria**

Round in cross section. Pointed at ends. Moves by wriggling. No eyes. (roundworms) Phylum **Nematoda**

Segmented and without a hard shell. May have a chitinised cuticle and jointed limbs. Phylum Arthropoda 14

12 Animal with hard shell and unsegmented body. Phylum Mollusca 13

13 Shell in two parts joined by a hinge. (mussels) Class **Bivalvia**

Shell in one part and either coiled or conical. Class **Gastropoda**– see Key 3

14 *Either* without jointed limbs, *or* with three pairs of jointed limbs, biting mouthparts (see Fig. 1) and no wings. Insect larvae –see Key 2

With jointed limbs, but not matching description above. 15

15 Body in three sections; head, thorax (three segments), abdomen (less than 11 segments). Three pairs of legs on thorax. (see Fig. 2) Class Insecta 16

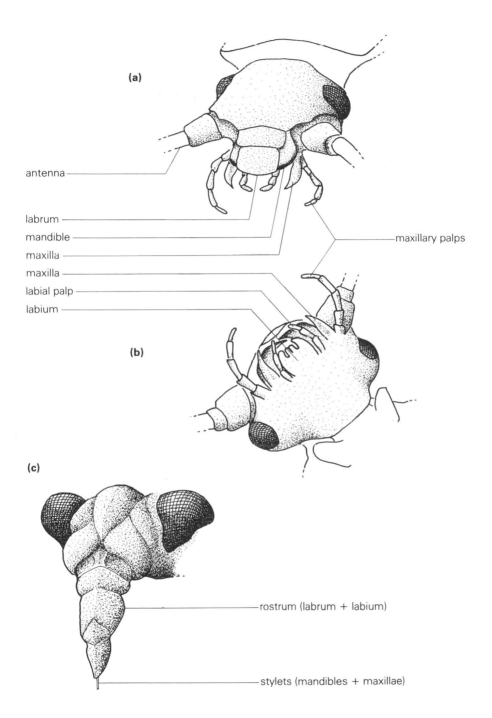

(a)

antenna

labrum

mandible

maxilla

maxilla

labial palp

labium

(b)

maxillary palps

(c)

rostrum (labrum + labium)

stylets (mandibles + maxillae)

Figure 1 Typical insect mouthparts; (a) & (b) biting type (stonefly larva) in dorsal and ventral views; (c) piercing and sucking type (water scorpion) in anterior view.

Body in two parts. Segmentation not obvious. Four pairs of legs.

Class Arachnida **24**

Division of body variable, but always more than four pairs of limbs.

Class Crustacea **25**

16 First pair of wings hard, forming wing cases. Biting mouthparts.

(water beetle) Order Coleoptera **17**

First pair of wings leathery or absent. Piercing and sucking mouthparts (see Fig. 1c). (water bug) Order Hemiptera **21**

17 Maxillary palps (see Fig. 1) longer than antennae, which are clubbed

Family **Hydrophilidae**

Maxillary palps shorter than antennae **18**

18 Small black beetle with middle and hind legs modified as short paddles. Swims rapidly, mainly on water surface.

(whirligig beetles) Family **Gyrinidae**

Middle and hind legs not short and paddle-shaped **19**

19 Slow-moving, walking beetle. Legs relatively long, last segment swollen and equipped with strong claws. No air bubble beneath the wing cases.

(riffle beetle) Family **Elminthidae**

Swimming beetle with fringe of hairs on hind legs. Air bubble trapped under wing cases. **20**

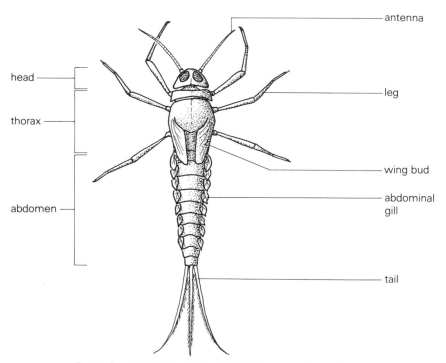

head

thorax

abdomen

antenna

leg

wing bud

abdominal gill

tail

Figure 2 Insect body parts (in this case a baetid mayfly larva).

20 Bases of hind legs covered by large plates, so joints between legs and thorax not visible. Family **Haliplidae**
Joints between hind legs and thorax not covered by plates. Family **Dytiscidae**

21 Living on water surface, antennae long and conspicuous. (**water measurer, water cricket** and **pond skaters**)
Living below surface. Antennae short and concealed. **22**

22 Breathing tube on end of abdomen. Moves by walking slowly. (**water scorpion** and **water stick insect**)
No breathing tube. Swims rapidly using long back legs. **23**

23 Swims ventral side uppermost. Mouthparts form long and pointed 'beak' below head (see Fig. 1c). (great water boatman) *Notonecta* **spp.**
Swims dorsal side uppermost. Beak short and blunt. (lesser water boatman) Family **Corixidae**

24 Like a terrestrial spider. At least 7 mm long. (water spider) *Argyroneta aquatica*
Less than 5 mm long and globular. Often brightly coloured. Swims rapidly. (water mite) Order **Acarina**

25 Usually at least 3 mm long. Subclass Malacostraca **28**
Usually less than 3 mm long. **26**

26 Bean-shaped. Entirely within hinged shell (called a carapace). (seed shrimp) Subclass **Ostracoda**
Not entirely enclosed in a carapace. **27**

27 Body laterally flattened. (water flea) Subclass **Cladocera**
Body pear-shaped. Often dark egg sacs attached to tail. Subclass **Copepoda**

28 Body dorsoventrally flattened. Like a wood-louse. (water louse) Order **Isopoda**, Genus *Asellus*
Body laterally flattened. Walks on one side, but sometimes swims. (freshwater shrimp) Order **Amphipoda**, Genus *Gammarus*
Body not flattened. Like a small lobster. (crayfish) Order **Decapoda**, *Astacus pallipes*

Key 2: a key to insect larvae common in fresh water

1 Without jointed legs. Order Diptera **2**
With three pairs of jointed legs on thorax. **7**

2 Thoracic segments fused together and wider than abdomen. **3**
Thoracic segments separate, not wider than abdomen. **4**

3 Larva transparent, except for two dark air sacs at either end. (phantom midge) *Chaoborus* **spp.**
Larva not transparent. No air sacs. (larvae of mosquitoes) Family **Culicidae**

4 Chitinised part of head small and incomplete or absent. Head can be retracted into thorax. **5**

Head completely chitinised and not retractile. **6**

5 End of abdomen with two dark-coloured spiracles (gridded openings of the tracheal system) set in a flat area bounded by six lobes. No stumpy, unjointed legs (prolegs) on abdomen.

(craneflies) Family **Tipulidae**, Subfamily **Tipulinae**

End of abdomen with spiracular plate bounded by fewer than six lobes or terminated in some other way. May have several pairs of prolegs on abdomen.

Suborders **Brachycera** and **Cyclorrapha**: other **Tipulinae**

6 Larva dumb-bell shaped with proleg behind the head. Circle of small hooks on end of abdomen. (blackflies) Family **Simuliidae**

Larva parallel-sided. **7**

7 Larva without prolegs. (biting midges) Family **Ceratopogonidae**

Larvae with prolegs, usually a pair behind the head and another pair at the end of the abdomen.

(non-biting midges) Family Chironomidae **8**

8 Antennae can be retracted into head.

(predatory chironomids) Subfamily **Tanypodinae**

Antennae not retractile. **(non-predatory chironomids)**

9 Wing buds present on last two thoracic segments (see Fig. 1). **10**

No wing buds. **14**

10 Three projections on end of abdomen, may be long 'tails' (see Fig. 2) or short points. **11**

Two long 'tails' on end of abdomen.(stoneflies) Order Plecoptera **13**

11 Lowest mouthpart (labium) modified to form extensible 'mask' (prey-catching organ, see Fig. 20). No gills on abdomen.

Order Odonata **12**

No 'mask'. Gills on some abdominal segments.

(mayflies) Order **Ephemeroptera**

12 Abdomen ends in three long 'tails'.

(damselflies) Suborder **Zygoptera**

Abdomen ends in inconspicuous points.

(dragonflies) Suborder **Anisoptera**

13 Upper mouthpart (labrum) more than twice as wide as long. Ventral part of tenth abdominal segment much shorter than the dorsal part (see Fig. 3a). **(predatory stoneflies)**

Labrum less than twice as wide as long. Ventral part of tenth abdominal segment not reduced (see Fig. 3b).

(non-predatory stoneflies)

14 Abdomen ends in two shortlegs, each bearing a short, stout hook. May be living in a case.

(caddis-flies) Order Trichoptera **15**

Not as above. **19**

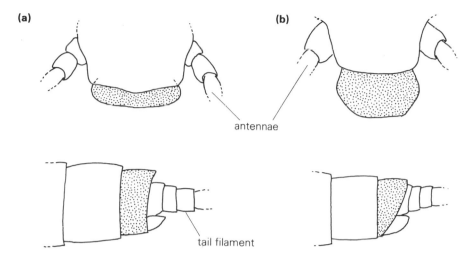

Figure 3 Labrum and abdomen of (a) predatory stonefly larva, and (b) non-predatory stone-fly larva. Labrum and tenth abdominal segment stippled.

15 Larva without a case.   **16**
    Larva within a case.   **18**
16 No tufted gills on abdomen.   Family **Polycentropodidae**
    With tufted gills on abdomen.   **17**
17 Gills attached to sides of abdomen. Thoracic segments 2 and 3 unchit-
    inised and soft. Often greenish in appearance.
                                                                    Family **Rhyacophilidae**
    Gills attached to underside of abdomen. Dorsal parts of thoracic
    segments 2 and 3 chitinised and hard.   Family **Hydropsychidae**
18 Very small larva (less than 7 mm long). Case purse- or flask-shaped,
    made of translucent, secreted material.   Family **Hydroptilidae**
    Case tortoise-shell shaped, made of small stones or grains of sand.
                                                                    Family **Glossosomatidae**
    Case construction various but shape generally tubular.
    Larva may be much longer than 7 mm.   **(remaining caddis-flies)**
19 Abdomen thin-skinned and soft.   **20**
    Abdomen heavily chitinised and hard.
                            (non-predatory larvae of beetles) Order Coleoptera 21
20 Abdomen ends in a single long translucent 'tail'. Translucent filaments
    along the sides of the abdomen.
                            (alder-flies) Order **Megaloptera**, Genus *Sialis*
    Abdomen ends in two 'tails', which may be very short.
                            (predatory larvae of beetles) Order **Coleoptera**
21 Single long tail on abdomen. Backward-pointing spines on thoracic
    and abdominal segments.   Family **Haliplidae**
    No long abdominal 'tail'. No backward-pointing spines.   **22**

22 Antennae more than twice as long as head.   Family **Helodidae**
   Antennae less than twice as long as head.
                              (riffle beetle larvae) Family **Elminthidae**

Key 3: a key to common freshwater gastropods

1 Snail with horny plate or operculum which can be used to close the
  shell aperture (see Fig. 4).   Subclass Prosobranchia **2**
  Snail without operculum.   Subclass Pulmonata **9**
2 Spire very low, scarcely protruding above last whorl (see Fig. 4).
  Aperture half-moon shaped. Shell usually patterned with paler oval
  blotches.   (nerite) *Theodoxus fluviatilis*
  Aperture not half-moon shaped. No pale markings.   **3**
3 Shell more than 30 mm high and usually with coloured bands on
  whorls.   **4**
  Shell less than 20 mm high and seldom banded.   **5**
4 Shell glossy. Umbilicus conspicuous (see Fig. 4).
                          (Lister's river snail) *Viviparus fasciatus*
  Shell dull. Umbilicus inconspicuous.
                              (river snail) *Viviparus viviparus*
5 Shell much taller than wide.   **6**
  Shell only just as tall as wide, or shorter.   **7**
6 Shell less than 6 mm tall. Operculum can be retracted beyond the
  aperture.   (Jenkins' spire shell) *Potamopyrgus jenkinsi*
  Shell more than 6 mm tall. Operculum not retractable beyond
  aperture.   *Bithynia* **spp.**

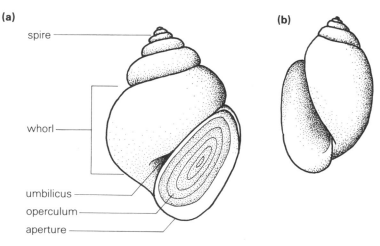

(a)

spire

whorl

umbilicus

operculum

aperture

(b)

Figure 4 The features of typical gastropods, as used in Key 3: (a) sinistral shell, open umbilicus; (b) dextral shell, closed umbilicus.

7   Shell flat and disc-like. Spire not rising above the level of last whorl.
         (flat valve snail) *Valvata cristata*
   Shell not flat. Spire rises above level of last whorl.   **8**

8   About as wide as tall. Width at least 5 mm.
         (valve snail) *Valvata piscinalis*
   About twice as wide as tall. Width less than 5 mm.
         Valvata macrostoma

9   Shell limpet-like, not clearly coiled.   **10**
   Shell clearly coiled.   **11**

10   Aperture at least twice as long as wide.
         (lake limpet) *Acroloxus lacustris*
   Aperture less than twice as long as wide.
         (river limpet) *Ancylus fluviatilis*

11   Shell flat. Spire not rising above level of last whorl.
         (ramshorns) *Planorbis* and *Segmentina* **spp.**
   Shell not flat. Spire projecting above last whorl.   **12**

12   Shell sinistral (see Fig. 4a).   **13**
   Shell dextral (see Fig. 4b).   **14**

13   Shell pointed. Height of aperture not more than half the height of shell.   (moss bladder snail) *Aplexa hypnorum*
   Shell blunt. Height of aperture scarcely more than half the height of shell.   (bladder snail) *Physa fontinalis*

14   Height of aperture much greater than half the height of shell.   **15**
   Height of aperture scarcely more than half the height of shell.   **16**

15   Aperture large with outer margin bent outwards. Spire points backwards when snail moves.   (ear pond snail) *Lymnaea auricularia*
   Outer margin not bent outwards. Spire points to side when snail moves.   (wandering snail) *Lymnaea peregra*

16   Large snail, shell up to 50 mm tall. Spire finely pointed.
         (great pond snail) *Lymnaea stagnalis*
   Shell up to 25 mm tall. Spire less acute.   **17**

17   Umbilicus closed by lip of shell. Square patterns marked by ridges on the whorls. Up to 25 mm tall.
         (marsh snail) *Lymnaea palustris*
   Umbilicus open. No square patterns on the shell. Up to 12 mm tall.
         (dwarf pond snail) *Lymnaea truncatula*

Abiotic environment

In many ways, fresh water is a demanding medium in which to live. Physical and chemical conditions can differ greatly between lakes and rivers and even between one lake or river and the next. Dramatic variations in conditions often occur within short distances in a lake or river, in both the horizontal and vertical planes. In any one place great fluctuations in abiotic conditions can occur from season to season or within as short a space of time as one day.

The physical and chemical factors which perhaps have most importance for freshwater life are temperature and the supply of oxygen. These two are inextricably linked; the warmer the water, the lower the amount of oxygen which remains in solution. In addition, the rate at which organisms deplete dissolved oxygen by respiring increases as the temperature increases.

Heat and oxygen are mainly supplied to the freshwater environment through the water surface. The water is heated by radiation from the sun and by conduction from the atmosphere. Water has a much greater specific heat than air; that is to say, much more energy, compared with air, is required to raise the temperature of water by a given amount, so the temperature of a water body changes much less rapidly than the air above it. In this way, aquatic animals are protected against the sudden changes in temperature to which terrestrial animals are exposed.

The arrangement of the molecules in water is such that water is most dense at $4\,^{\circ}$C. This means that ponds freeze first at the surface, and the warmer, denser water deeper in the pond is protected from freezing by a cooler layer floating on top. It is unusual for a pond, except a very small one, to freeze solid.

Oxygen enters by diffusion from the atmosphere, although the photosynthesis of water plants provides an additional source. In fast-flowing rivers and streams oxygen is rarely in short supply, since the turbulent flow ensures that the water is rapidly aerated. Instead the animals and plants that live in rivers have the problem of resisting the current, which tends to wash them downstream and ultimately out to sea.

In chemical terms, fresh water is a complex cocktail of many dissolved substances. Sodium, calcium and bicarbonate ions are usually the most abundant inorganic substances, but phosphate and nitrate, generally much less abundant, have importance as plant nutrients. Unlike the sea, the concentrations of all these substances in fresh water are often very low, so complex apparatus and considerable skill are needed to measure them accurately.

A chemical factor which is more easily measured is the concentration of hydrogen ions, or pH. Acid waters have a high concentration of hydrogen ions and, since the pH scale is inversely related to hydrogen ion concentration, a low pH. Conversely, alkaline waters have a low hydrogen ion concentration and a high pH. Hydrogen ion concentrations in the sea rarely depart much from pH 7, the neutral value, because the dissolved salts act as a buffering system. However, natural pH values in fresh water can be as low as 4 or as high as 8. Since pH is measured on a $\log_{10}$ scale, the range between pH 4 and pH 8 represents a 10 000-fold change in hydrogen ion concentration.

Chemical conditions in fresh water depend principally on the nature of the rock and soil in the land from which the water has drained, called the catchment area. Water draining from an intensively farmed catchment containing rock which is easily eroded, such as limestone, will be rich in dissolved minerals and strongly buffered at about pH 8. On the other hand, water draining from a moorland catchment on a hard rock such as granite will be poor in dissolved substances, weakly buffered and variable in pH, usually less than pH 7.

Exercise 1: physical and chemical factors in ponds

Anything from a garden pond to a small lake or even a stagnant canal or river backwater will suffice as a 'pond'. The value of the exercise can be increased if two or more ponds differing in size or other physical attributes can be compared.

Safety note: this exercise should be supervised at all times. A wading stick should be used to probe for the bottom in unfamiliar areas. Life-jackets are a wise precaution. This warning applies equally to other exercises involving wading.

Aims

(1)   To investigate the physical and chemical conditions in a pond.
(2)   To record the changes in abiotic conditions over a 24 hour period.

Materials

Measuring tape; metre rule; water-sampling bottle (see Fig. 5); thermometer; meters for oxygen, pH and conductivity (or alternative laboratory methods, see below); collecting kit (see Appendix 1); graph paper.

Time

3 h (aim 1); 24 h (aim 2).

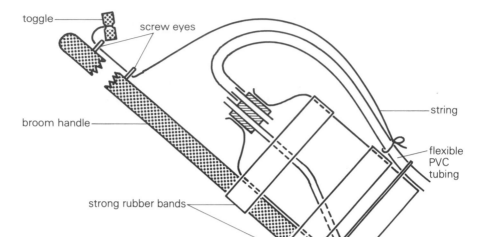

Figure 5 A simple water-sampling bottle.

Method

Draw a sketch map of the pond to scale, marking the depth at appropriate intervals (close together where the depth changes rapidly, further apart where it does not), and plot the distribution of water plants. Mark the position of any streams feeding or draining the pond. With large ponds, an outline taken from the Ordnance Survey 1 : 25 000 map can help. Try to find out as much as you can about the rocks in the catchment area.

Use the water-sampling bottle to collect water samples in the deepest part of the pond that you can reach, from just below the surface, about mid-depth and just above the bottom. Once the bottle is in position at the right depth, take a sample by gently pulling on the string to raise the end of the flexible tube above the bung. When bubbles stop appearing on the surface the bottle is full. The sample can be siphoned out of the bottle by attaching the flexible tube to the long glass tube instead of the short one. Start the siphon by gently blowing into the short tube. Take additional samples from the inlet and outlet streams close to where they join the pond. If there is a bed of submerged water plants take a sample from between the plants.

Measure the temperature, pH and oxygen concentration of the samples as soon as possible after they are taken and record the results. If you do not have meters, use test papers for pH and the Winkler titration method for oxygen (see Appendix 2). Read the concentration of oxygen in saturated water at the temperature of your samples from Figure 6. Calculate the percentage oxygen saturation of your samples using the following formula:

$$\% \text{ saturation} = \frac{\text{sample concentration}}{\text{saturation value}} \times 100$$

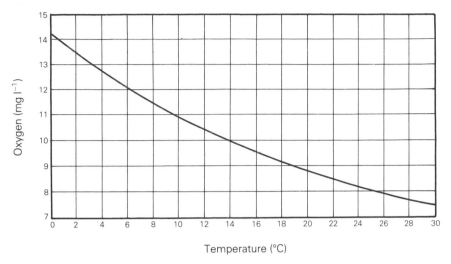

Figure 6 The relationship between the oxygen content and the temperature of fresh water saturated with air at normal pressure (1 bar).

Conductivity is an index of the total concentration of ions in the water and can be measured later. Instead of a meter you can measure dissolved solids by evaporating a known volume of filtered water (at least 100 cm³) to dryness in a weighed container and reweighing. Use an accurate balance and take care that the water does not spurt while it is being boiled.

Collect animals from different depths and habitat types in the pond; make careful notes of the habitat type from which each sample came. Identify them using the keys and plot their positions on the map. Estimates of numbers are not required here, but if an attempt is made to sample the animals and plants of a weed bed quantitatively (see Exercise 6), then the data can be used in Exercise 20. In this case, preserve the animals in 70 % ethanol and dry the plants for later examination.

If you intend to follow the changes in abiotic conditions over 24 h, take water samples at, say, two-hourly intervals and plot the results against time on graph paper. Take particular care when sampling in the dark.

Questions

(1) Why is it desirable to:
  (a) have a layer of paraffin in the water-sampling bottle?
  (b) measure oxygen concentrations as soon as possible after collecting the samples?
(2) Explain any differences that you have found in abiotic conditions between different parts of the pond.
(3) Which habitat type contains the greatest number of species (diversity) and which least? List the common species in each habitat. How does the distribution of animals relate to abiotic conditions?

(4) Explain the daily patterns in abiotic conditions that you have observed and comment on the reliability of the day's results as an indicator of the 'typical' conditions.

(5) In what ways would you expect your results to differ if the exercise had been conducted in a stream? If possible compare the results of this exercise with those of Exercise 2.

Exercise 2: the abiotic environment of a stream

Aim

To study the changes in current speed, depth, bottom type and other physiochemical conditions along a reach of stream.

Materials

Measuring tape; metre rule; an orange; stopwatch; water-sampling bottle (see Fig. 5); thermometer; meters or laboratory equipment for measuring pH, oxygen, conductivity (see Exercise 1); collecting kit (see Appendix 1); graph paper.

Time

3 h.

Method

This exercise is best conducted on a stream which has a gravelly bottom, is at least 2 m wide and is shallow enough to wade safely. Select a section of stream that includes a bend and a straight stretch. This will generally contain a part where the water is shallow and fast flowing (a riffle) and a part in which the water is deeper and slower flowing (a pool).

Make an outline map of the stream and select places at which the stream cross section will be investigated. If only two cross sections are being drawn, make one in the middle of the bend and the other in the middle of the straight section. Mark the positions of the cross sections on the map. At each place measure the width of the stream and measure stream depth at five or more points across the stream. At each point across the stream where the depth was measured record the current speed by measuring the length of time taken by an orange to float a measured distance. Use a pond net or a bucket to collect a sample of the gravel at each of the points where the current speed was measured. Collect a water sample from each point and measure its temperature, pH and oxygen concentration (more detailed instructions for water sampling are given in Exercise 1).

One way to collect benthic animals from streams is to hold a pond net vertically so that the bottom edge of the frame is touching the bottom of the stream. Vigorous kicking at the gravel upstream of the mouth of the net will dislodge the animals, and they are then washed into the net by the current. This is called 'kick' sampling. Collect 'kick' samples from the points near where the current speed was measured.

Calculate the average size of the gravel in each sample from measurements of 30 particles selected at random. Identify the animals in the samples and count the number of each type. Draw cross sections of the stream to scale on graph paper. If the stream is wide and shallow it may help to draw the depth and width to different scales. Mark the current speeds on the cross sections. Show the number of each type of animal as kite diagrams or histograms drawn below the stream cross sections.

Questions

(1) Describe the relationship between current speed and the average size of the gravel (plot current speed against gravel size on graph paper). Explain your findings.

(2) Explain the patterns in temperature, pH and oxygen concentration that you have found along the stream.

(3) List the dominant organisms and typical abiotic conditions in:
  (a) riffles;
  (b) pools;
  (c) stream margins.
  Which part of the river has the most diverse fauna?

(4) Suggest which abiotic factors may be affecting the distribution of animals in the river. Design experiments to test your hypothesis.

(5) How might abiotic conditions differ in a pond? If possible compare your results with those of Exercise 1.

Adaptations

There are very few groups of organisms which are found only in fresh water. The majority of freshwater organisms appear to have been colonists from other environments in which they still have many close relatives. In most cases the exact timescale of colonisation is unknown, but it is likely that it was a gradual process in which species evolved tolerance of freshwater conditions over many generations.

Freshwater crustaceans, for instance, originated in the sea; the first colonists were probably species that were already adapted to the estuarine environment. The insects in fresh water came from terrestrial sources; many are related to species found in soil and other damp places. The problems faced by the colonists in adapting to their new surroundings depended on the environment from which they had come.

Colonists from the sea had to adapt to the low concentration of salts in fresh water. In a medium which is more dilute than body fluids, and therefore has a higher osmotic potential, water enters cells by osmosis. Salts that are lost in excretion or by diffusion are difficult to replace. Most marine colonists solved this problem by a combination of increased tolerance of more dilute body fluids, active uptake of salts from the water, and production of a urine with a lower concentration of salts than their body fluids (i.e. hypotonic). The problem was not so acute for the colonists from terrestrial environments since life on dry land required a relatively impermeable body surface. This impermeable cuticle, necessary to reduce water loss in air, pre-adapted the terrestrial colonists to life in fresh water, in which the problem was to keep water out and salts in.

Oxygen is always present in much higher concentrations in air than in water. The difficulties of respiring in water were familiar to the marine colonists but new to the invaders from dry land. Some avoided the problem by keeping a connection to the water surface through an air-filled tube or snorkel. Others evolved ways of carrying a store of air, which has to be renewed by visits to the surface, much like the aqualung used by human scuba divers. Some insects solved the problem by adapting their tracheal respiration system to function as a gill, taking up dissolved oxygen from the water either by diffusion through areas of thin cuticle (a 'biological' gill), or across the surface of a film of air adhering to the body (a 'physical' gill). As well as respiratory problems, life in fresh water presented new opportunities for feeding and dangers from new predators; we investigate adaptations to these in the following exercises.

Exercise 3: how a freshwater organism respires: a study of corixids

Background

Corixids are often abundant in ponds, the edges of lakes and in very slow-flowing rivers (see Fig. 7). They are bugs (Order Hemiptera) and are commonly called lesser water boatmen. Unlike most bugs they are not exclusively carnivorous but feed mainly on microscopic plants and decaying organic matter. Corixids carry a store of air attached to fine hairs on the surface of their bodies.

Aim

To study the respiration of corixids in water containing different concentrations of dissolved oxygen.

Materials

Conical flasks (250 cm³); small quantity of 'Netlon' plastic mesh; stopwatch or clock; aerator pump.

Time

1 h for the exercise plus ½ h preparation the previous day.

Figure 7 Corixid water bugs surfacing to renew their stores of air. Note that the right-hand bug has broken through the water surface film (photo by R. A. Moore).

Method

This experiment requires 500 cm³ of cool, deoxygenated water. Boil this quantity for 10 minutes and allow to cool overnight in stoppered containers which have been filled to the top to exclude air. This procedure should reduce the dissolved oxygen to about 30 % of the saturation value at room temperature. Immediately before the exercise aerate a further 1.5 l of water for at least 10 min.

Fill one beaker with deoxygenated water and one with water which is saturated with air. Prepare 1 l of water which is 65% saturated by mixing half litres of deoxygenated water and saturated water. The concentration of oxygen in each of the three beakers can be estimated by referring to Figure 6 or by direct measurement (see Exercise 1). If bottled nitrogen or oxygen is available, the exercise can usefully be extended to include extra flasks containing water through which either gas has been bubbled for about 10 min. Gassing with nitrogen reduces dissolved oxygen to almost zero, while gassing with oxygen displaces nitrogen and increases oxygen concentrations to far above the air saturation level. Place a small piece of plastic mesh at the bottom of each beaker, weighted with a coin if necessary.

Gently place five corixids in each beaker. Try to ensure that the animals in each of the beakers are similar in size. The species name is not important here, but can be found from the key by Macan (1965; see Bibliography).

Sit quietly beside the beakers and do not make any sudden movements that might disturb the animals. Wait five minutes (to allow the animals to become accustomed to their surroundings), and then count the number of times in the next 45 min that the animals return to the surface to replenish their supply of air. Do not count trips that stop short of the surface. Observe and record any movements that the corixids make while resting on the mesh.

Questions

(1) Account for differences between the number of trips made by the animals in each beaker. What does this tell you about the method of respiration used by corixids?

(2) Where does a corixid carry its air store? Suggest how the air is made to 'stick' to the surface of the cuticle.

(3) Describe the movements made by the corixids whilst resting and suggest a function for them. Design an experiment to test your hypothesis.

(4) Discuss the advantages and disadvantages of respiring in this way in comparison with the use of a snorkel or a biological gill.

(5) What purpose does the plastic mesh serve? Why would living plants not be a suitable substitute in this experiment?

Exercise 4: an investigation of case building by caddis larvae

Background

Caddis flies are closely related to butterflies and moths but their larvae are almost always found in fresh water or at least in very damp places. The larvae of the majority of species construct portable tube-shaped cases in which they reside both as larvae and, later, pupae (see Fig. 8). Even those that do not make cases out of foreign matter often spin temporary retreats made of a sort of silk. Unfortunately, it is not easy to identify some species of caddis larvae, but it is safe to assume that larvae that look different, or that live in the same place but in different sorts of cases, are different species. Caddis larvae may be found in almost any fresh water, although some practice may be necessary to spot their cases amongst the materials from which they are made!

Aim

To investigate the structure and function of caddis cases.

Materials

Collecting kit (see Appendix 1); blunt needle or pin; fine forceps; aquarium and aerator pump or small, squat dishes (e.g. evaporating dishes or cottage cheese cartons); suspension of fine carmine particles in water; Pasteur pipette.

Time

1 h for simple laboratory work; 3 h if including fieldwork; several days or more if investigating case rebuilding or growth.

Method

Collect caddis larvae from a convenient stream or pond. Record the type of habitat from which the larvae were collected (e.g. weed bed, gravel) and collect suitable case-building materials from the same spot. In the laboratory the larvae may be kept together in an aerated aquarium or individually in small dishes which need not be aerated.

Gently release a small amount of carmine suspension at the front of the case just above the head of a larva. Observe the fate of the carmine. Release a small amount of carmine at the end of the case and observe the result.

Remove the larva from its case by probing gently with a blunt pin from the rear. Measure the length of the larva, the length of its case and the width

Figure 8 A selection of caddis-fly larval cases. Note the diversity of materials and styles of construction (photo by R. A. Moore).

of the case at the front and rear ends. Repeat this procedure for a series of larvae of the same species. Study the method of construction of the case, if necessary by partly dismantling the case with fine forceps. Record the sorts of material used in the case and the sizes of the particles. These can be accurately measured if a microscope with a mechanical stage or an eyepiece graticule is available, otherwise judge by eye whether the particles are all the same size or if they vary, perhaps along the length of the case.

If a small part of the front edge of an occupied case is removed its occupant will start to repair the damage. By providing suitable materials the rebuilding behaviour can easily be observed. It is necessary to set up dishes containing water and the building materials about two hours in advance and then to stir the dishes immediately before rebuilding commences so that the small bubbles which usually collect in the dishes are dispersed. Further suggestions for projects involving caddis cases are provided by Hansell and Aitken (1977; see Bibliography).

Questions

(1) What could be the function of water movement through the case? Design an experiment to test your hypothesis.
(2) What protects a caddis larva from attack through the rear end of the case?
(3) Are the cases of large and small larvae similar in size and construction? What can you infer about changes in cases as larvae grow?
(4) How does a larva select suitable case materials? How are the pieces stuck together?
(5) What functions might the case serve? How would you test your hypothesis?

Exercise 5: a study of attachment and feeding by blackfly larvae

Background

Larvae and pupae of blackflies (*Simulium* spp.) are found exclusively in flowing water, from mountain torrents to gently moving lowland streams. In clear, fast-flowing streams larvae may often be seen attached in large numbers to the upper surfaces of stones.

Blackfly larvae are exquisitely adapted for filter feeding. Food particles passing in the current are caught on the extended head fans (see Fig. 9) and transferred to the mouth. Until recently it was a puzzle how such a relatively coarse filter could catch particles as small as individual bacteria (1–2 μm diameter), but now it appears that they are caught in a thin layer of mucus spread on the head fans. When they pupate, the larvae spin a slipper-shaped

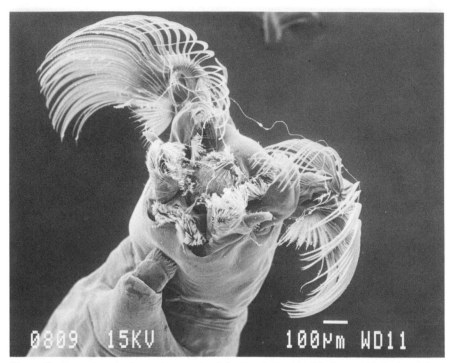

Figure 9 Scanning electron micrograph of the head of a blackfly larva with its head fans extended. The fine threads are fragments of the silk used by the larva for attachment to the surfaces of stones.

silken case for protection and breathe through antler-like respiratory horns which are attached to their tracheal systems.

The adults are biting flies, which in Britain principally attack cattle, but in some tropical areas blackflies transmit a serious human disease called river blindness.

Aims

To study the adaptations of blackfly larvae to filter feeding.

Materials

Aquarium; piece of glass or plastic; aerator pump and air stone; mounted needle; baker's yeast; 95 % alcohol; cavity slide and coverslip; microscope; hand lens; pipette.

Time

3 h if including fieldwork; 1 h if material is collected and set up in advance.

Method

The simplest way to collect blackfly larvae is to pick up the stones or river weeds to which they are attached (pupae may be found firmly attached to stones or weeds in the same area, but probably not in positions exposed to the full force of the current). Make notes of the places in which you found the larvae. Keep the larvae in cool, well-aerated or damp moss while they are in transit to the laboratory. Set the aquarium and air stone up as shown in Figure 10 and introduce the larvae.

Leave the larvae to settle for at least 1 h and then record the distribution of the larvae in the tank. Select a larva which is actively feeding (head fans extended) and observe with a hand lens. Count the number of times that the head fans are retracted in 1 min. Add a small amount of yeast suspension to the tank and repeat the observation. Using a mounted needle, gently detach a larva and observe what happens. Place a ruler or similar obstruction in the tank so that the flow of water past some of the larvae is changed. Observe their reaction. Finally, turn off the aerator pump and observe the result.

Pipette some of the larvae from the tank and preserve them in alcohol. Mount one of the larvae, preferably one with the head fans at least partially extended, on a microscope slide and examine under the microscope. If the larva is a large one it may be necessary to support the coverslip on small pieces of Plasticine.

Questions

(1) From your observations, suggest how the larvae are adapted for attachment in fast currents.
(2) Why do the larvae select certain sites for attachment in the river and the laboratory tank but reject others?
(3) Blackfly larvae are often very abundant in rivers immediately below a reservoir or lake. Why?

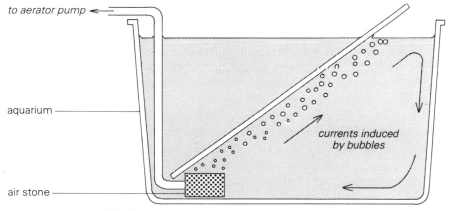

to aerator pump

aquarium

currents induced by bubbles

air stone

Figure 10 Experimental tank for the blackfly feeding exercise.

Populations

While other biologists may be concerned with the properties of cells, tissues or even whole organisms, ecologists are interested in the processes that control the distribution and abundance of individual organisms or groups of organisms. A group composed of individuals of one species living together in a particular place, perhaps a pond, is called a population.

The number of individuals in a population is known as the population size (not to be confused with population density – see Exercise 6). Reproduction by individuals in the population and immigration from nearby areas cause populations to increase in size. Death and emigration cause populations to decrease in size. Of course if a population exists in an area for a long time without greatly changing in size, as is often the case, then the increases and decreases must balance. As a first step towards understanding the factors that influence the changes in population size (population dynamics) it is necessary to be able to make a reasonably accurate census of populations. In this chapter we shall look at several ways of estimating population size and also investigate the factors affecting the dynamics of laboratory populations of freshwater invertebrates.

Exercise 6: investigating the abundance of a population of freshwater invertebrates

Background

It is rarely feasible to count each individual in an entire population of invertebrates. Instead, a sample is taken from the population in such a way that the number of individuals in the sample is related to the number of individuals per unit area in the population (called population density). These are called quantitative samples, as opposed to qualitative samples which merely show which species are present and do not give a reliable indication of their density.

There are two main methods of quantitative sampling suitable for freshwater environments. One involves using the same amount of effort to take each sample and then counting the number of individuals yielded by that effort. If samples are taken in the same way in several places or at different times, then the number caught on each occasion should be related to the density of the population. A method based on sampling effort can show differences in population density from place to place or from time to time, but it cannot give an absolute figure for the density at any one place or time.

In the second type of quantitative sampling the aim is to catch all the animals of a particular species that live within a known area of the habitat. This method yields an absolute figure for population density which can be used to calculate the population size if the total area occupied by the population is known.

Both types of quantitative sampling, effort-based and area-based, are described in this exercise. Methods are described which are suitable for ponds or for rivers, so that this exercise could be carried out in almost any freshwater environment.

Aim

To use an effort-based or area-based sampling method to estimate the size of a population of freshwater invertebrates and to find its spatial distribution pattern.

Materials

(a) Ponds; collecting kit (Appendix 1), surveyors tape or large scale map.
(b) Rivers; collecting kit, metal quadrat (a square frame of stiff wire will do), surveyors tape.

Time

1 h in the field plus 2 h in the laboratory.

Method

Aim to estimate the size of a population within a uniform area of habitat; this might be a whole pond, a riffle or pool on a river or just a single weed bed. Whichever method of sampling is used you should take a sample consisting of at least ten sampling units. A sampling unit is the collection of animals that is made each time the chosen sampling technique is used. You should take the sampling units from randomly selected locations within the chosen habitat.

In weed beds effort-based samples can be taken by sweeping the pond net through the weeds in a standard way for a fixed number of times. Trial and error will show how many standard net sweeps are necessary to produce a reasonable catch of animals in each sampling unit.

'Kick' sampling is an effort-based sampling method suitable for rivers. This sampling method is described in Exercise 2, but when using 'kick' sampling in a quantitative way you must be sure to use the same number of kicks with constant enthusiasm for each sampling unit!

Area-based samples are difficult to take in very thick weed beds but in places where the weed is sparse it is possible to collect the bottom-dwelling organisms in a known area. In rivers or ponds this can be done by pushing the pond net through the sediment for a known distance which should be less than the length of the net bag. The net should be pushed up stream in a river. If this distance is multiplied by the width of the mouth of the net then the area from which the sampling unit came can be calculated.

In stony rivers, pushing the net is difficult and it may become damaged. An alternative technique is to lay a quadrat on the river bed just up stream of the mouth of the net. Lift the larger stones from inside the quadrat and wash off any clinging animals in front of the net mouth, so that the current carries the animals into the net. Include stones which overlap two of the quadrat edges and exclude stones overlapping the other edges. When all the large stones have been removed stir the finer gravel vigorously so that the remaining animals are dislodged and washed into the net.

Finally, measure the area of habitat occupied by your population.

In the laboratory, choose one species for study and pick out and count all the individuals of that species in each sampling unit. Calculate the mean number of individuals per unit in the sample ($\bar{x}$) and the sample variance (s^2). The variance gives a measure of the amount of variation in counts that occurs between sampling units; if all sampling units contained the same number then the variance would be zero. To find the variance first subtract the sample mean from the number in each sampling unit. Next square each

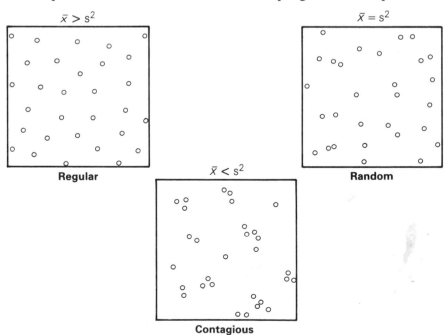

Figure 11 Three spatial dispersion patterns and their variance to mean relationships.

of these differences. The variance is the sum of all the squares divided by one less than the number of sampling units (n) i.e.:

$$s^2 = \frac{\sum(x - \bar{x})^2}{n - 1}$$

In a sample of invertebrates, the size of the variance in relation to the mean reflects the dispersion pattern of the individuals or the way in which they are distributed over the bottom of the stream. Some possible dispersion patterns and their variance to mean ratios are shown in Figure 11. These relationships between the variances and means of the dispersion patterns are only exact for extremely large samples. For samples which consist of between 10 and 30 units, it is reasonable to conclude that the population has a contagious (or clumped) pattern if the variance is more than twice the mean, and a regular pattern if the variance is less than one third of the mean. Variance to mean ratios that fall between these limits are likely to arise from a sample of a randomly dispersed population.

Questions

(1) If you have used an area-based sampling method, calculate the population size in your chosen habitat. The answer may surprise you!
(2) Discuss the possible sources of error in your estimate of population size. Why might a single census such as this give a false impression of the size of a population in a pond or stream?
(3) Describe the dispersion pattern of your chosen population. Discuss how the dispersion pattern might affect the number of sampling units required to give an accurate indication of population density.
(4) How would the size of the quadrat affect your ability to detect different dispersion patterns?
(5) Suggest some aspects of the biology of animals which might lead them to have regular, random or contagious dispersion patterns.

Exercise 7: estimating the size of a stickleback population

Background

If animals can be marked in such a way that the same animals can be recognised if they are caught again, then the size of the population can be estimated by a method quite different from those in Exercise 6. In this method a sample of animals is caught and each individual marked. Some time later, another sample is collected and the size of the population calculated from the proportion of marked animals in the second sample. This method is known as the Lincoln or Petersen index.

Sticklebacks are such familiar fish that they scarcely need introduction. Two species are found in fresh water in Britain, the three-spined, *Gasterosteus aculeatus*, and the ten-spined, *Pungitius pungitius* (see Fig. 12). Fortunately, the two species can be distinguished by simply counting the number of spines along the back of the fish, although the so-called 'ten-spined' can have as many as twelve or as few as seven! Three-spined sticklebacks are particularly suitable animals for this exercise because they can be marked by clipping one of the two pelvic spines. This technique is painless and harmless to the fish.

Aim

To mark and recapture individuals of a stickleback population and to estimate population size using the Lincoln or Petersen index.

Materials

Pond nets; pair of strong, small scissors or nail clippers; two buckets.

Time

1 h on each of two occasions about one week apart.

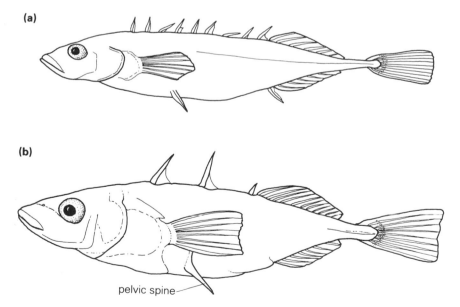

Figure 12 British freshwater sticklebacks: (a) ten-spined, *Pungitius pungitius*; (b) three-spined, *Gasterosteous aculeatus*.

Method

Use pond nets to collect sticklebacks from a pond, ditch, river backwater or canal and keep the captured animals in one of the buckets. Mark the captured animals by clipping off about half the length of one of the two pelvic spines. Decide beforehand which spine is to be clipped and mark all the fish in the same way. Handle the fish gently with wet hands so as not to damage them. Put the marked fish into the second bucket, counting them as you go, and when all the fish have been marked gently return them to the pond.

On the second visit try to capture about as many fish as were caught on the first visit. Count the fish as they are caught, inspect each one and record the number that are marked. Put fish that have been inspected into the bucket. When all the fish have been inspected return the contents of the bucket to the pond.

Calculate the number of sticklebacks in the pond using the following formula:

$$P = \frac{mt}{r}$$

where P is the estimate of the population size, m is the number of fish marked on the first visit, t is the total number of fish captured on the second visit and r is the number of marked fish that were recaptured.

Questions

(1) How would immigration to the pond or emigration from the pond between the first and second visits affect the population estimate?
(2) How would the estimate be affected if marked fish were more likely to die between the visits than unmarked fish?

Exercise 8: population growth in water fleas

Background

Water fleas are small crustaceans belonging to the Order Cladocera. Most freshwater cladocerans are members of the zooplankton and they are often very abundant in ponds and lakes. 'Zooplankton' is the term used to describe the small animals that live suspended in the open water, often quite close to the surface. The term is rather a loose one; in practice it means everything that does not swim fast enough to avoid a plankton net!

Water fleas feed by filtering small particles from the surrounding water. By beating their legs back and forth, they generate a feeding current from which they strain the particles of food (compare this method with the filter feeding of blackfly larvae in Exercise 5). The food particles are single-celled plants (planktonic algae) and fragments of dead organic matter (detritus), together with associated bacteria.

For much of the year many water fleas reproduce asexually (partheno-genesis). Parthenogenetic females lay eggs from which hatch more par-thenogenetic females; usually the eggs are carried in a special brood pouch until they hatch. Males are only produced in the autumn and they mate with females to produce sexual eggs which remain dormant in the mud until spring.

Aim

To study the growth rate of cladocerans in laboratory culture.

Materials

Enough 250 cm³ beakers to give four to each group of students; small quantity of dried yeast; Pasteur pipettes; fine mesh pond net (<250 μm); filtering apparatus; autoclave or pressure cooker; plastic jerrican; aquarium air pump and air stone; balance to weigh 5 mg.

Time

Either 1 h to set up cultures plus ½ h at about four-day intervals over three weeks, or, if the cultures can be set up at intervals in advance of the class, 2 h to examine the cultures.

Methods

Use the pond net to collect water fleas from a pond. Fill the jerrican from the same pond.

Filter enough water to provide each group with 1 l. This process can be speeded up if a Buchner-type vacuum filtration apparatus is available. Autoclave the water at 10^5 Pa (15 lb in^{-2}) pressure for 15 minutes in flasks loosely plugged with cotton wool. Allow the water to cool to room temperature, aerate for 15 minutes and allow to stand for 30 minutes.

Each group should set up three culture flasks containing 150 cm³ of filtered, autoclaved, aerated water and one flask containing 75 cm³. Add yeast to each of the three 150 cm³ culture flasks at doses of 5, 10 and 20 mg, and add 10 mg to the 75 cm³ flask. The first three flasks contain different amounts of food in the same volume of water, whilst the last flask provides the same amount of food as the second flask but in half the volume of water.

Pick out individuals of one species of water flea using the Pasteur pipette and add ten individuals to each culture flask. The name of the species is not important here, but can be found from the key by Scourfield and Harding (1958; see Bibliography).

Examine the cultures at intervals of about four days, transferring the water fleas to a new culture flask set up as before and counting them as they are transferred. This process requires a further supply of treated pond water. Plot the number of water fleas against time from the start of the culture for each of the four culture flasks. If a number of groups have been running cultures their results for each food level or water volume could be combined and averages calculated.

Questions

(1)   How does concentration of food affect the growth rate of the cultures and, if the cultures reach an equilibrium, the final population density?
(2)   Explain how the amount of living space (water volume) affects the growth of the cultures.

Producers

The food supply for animals in freshwater environments comes from two main sources. Some arrives in the form of organic detritus which falls into the pond or stream. This is material which has been produced mainly by terrestrial plants, although animal remains can sometimes be quite significant. The other main source of food is that produced by organisms living within the freshwater environment. It is eaten either in the living state or after it has died and started to decay. It is the production of material by green plants living within ponds and streams that we shall be investigating in this chapter.

All green plants require access to a number of resources in order to survive and grow. Sunlight, water and inorganic carbon are required to enable photosynthetic production of carbohydrates. Oxygen is required for normal respiratory processes but this is also a by-product of photosynthesis. Certain additional inorganic nutrients, notably phosphates and nitrates, are necessary for the growth of new tissue. Terrestrial plants acquire carbon in the form of carbon dioxide from the atmosphere. They need roots in order to absorb nutrients and water from the soil, and they have stems to support the leaves in a position in which they can absorb light. When the productivities of terrestrial plant communities living in similar latitudes are compared, the major differences are usually due to the availability of water. Thus plant productivity in tropical rain forests is very high, whereas productivity in tropical deserts is always low.

Fully submerged aquatic plants rely on carbon dissolved in the water, either as carbon dioxide or the bicarbonate ion. Because they live immersed in a weak solution of nutrient salts, many aquatic plants do not require a root system. The medium also provides buoyant support for their tissues, so many aquatic plants function quite successfully without stems. Thus the smallest aquatic plants, the algae of the phytoplankton, are amongst the simplest structures in the plant kingdom.

Needless to say, aquatic plants rarely suffer from shortage of water. But aquatic plants which do not have roots in the sediment and therefore rely on absorbing mineral salts from the water, e.g. planktonic algae, are frequently limited by nutrient shortage. The critical nutrient is often phosphorus, in the form of phosphate.

A second factor that limits growth of aquatic plants is light. Plants with leaves below the water surface suffer reduced illumination because light is absorbed by the water above and also by any dissolved or suspended material within the water. The so-called emergent plants of marshes and reed beds, which have roots in the mud and leaves above the water surface, are rarely limited by nutrient supply or by light. So plant communities at the margins of aquatic habitats are among the most productive in the world.

Exercise 9: measuring production of phytoplankton

Background

Since planktonic algae are the main food source for animal communities in many freshwater environments, it is useful to be able to measure algal productivity, or the extent to which inorganic carbon is converted to organic carbon in photosynthesis. This allows the factors that restrict or enhance productivity to be investigated. Gross productivity is the total amount of inorganic carbon 'fixed' in organic compounds, whilst net productivity is the amount that is left after the metabolic needs of the algae have been met. In effect, net productivity is the extent of production of consumable plant tissue, so it is often the more useful quantity.

One way of estimating net productivity would be to measure the amount of new tissue as it is produced, perhaps by weighing or by counting cells. However, this method is tedious, frequently impractical and often inaccurate.

An alternative method uses changes in dissolved oxygen as an indirect measure of productivity. Algae that are enclosed in a transparent container and exposed to light, produce oxygen when they photosynthesise and consume oxygen when they respire. Any overall increase in the concentration of dissolved oxygen in the container indicates that the carbon fixed in photosynthesis is in excess of the amount required to maintain algal metabolism; it is therefore a measure of net productivity.

The extent of algal respiration can be found by measuring the change in dissolved oxygen which occurs when algae are incubated in a sealed container in the dark. Without light the algae continue to respire but are unable to photosynthesise. By adding the oxygen used in respiration to the increase in oxygen concentration that occurs in the light, a measure of gross productivity can be obtained. This is the basis of the 'light and dark bottle' method of measuring plant productivity.

Aim

To measure the productivity of phytoplankton using the light and dark bottle technique.

Materials

250 cm^3 reagent bottles with ground glass stoppers; household aluminium foil or black polythene; insulating tape; equipment for incubating samples (see Method); source of phosphate ions; equipment for measuring oxygen concentration (see Appendix 2).

Time

Two periods of 1 h, 2–4 h apart.

Method

Divide the reagent bottles into three equal groups. Cover the outside surfaces of the bottles of one group with aluminium foil or black polythene, holding it in place with insulation tape. Be sure that there are no holes through which light can pass; remember to cover the stoppers! The remaining two groups are left uncovered.

Taking the bottles three at a time, one from each of the groups, fill them with pond water containing phytoplankton. If algae are scarce in the water the bottles can be 'seeded' with cultured material. Make sure that the bottles are filled to the top and that there are no air bubbles trapped inside. Put identifying marks on each set of three bottles.

Immediately after filling the bottles, take one uncovered bottle from each set of three and measure the oxygen concentration. This is the initial oxygen reading. The remaining pairs of bottles (one covered 'dark' bottle and one uncovered 'light' bottle) should be incubated in the pond. Suspend the bottles using the apparatus shown in Figure 13, with each pair at a different depth. Alternatively, the bottles could be incubated in water baths in the laboratory, although it may be necessary to provide additional lighting. A simple measure of relative light intensity in the laboratory can be obtained using a photographic light meter in an incident light mode.

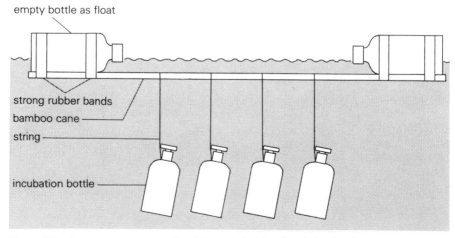

Figure 13 Apparatus for suspending light and dark bottles during incubation in a pond.

The optimal incubation time depends on temperature, light intensity, nutrient concentration and abundance of phytoplankton. In reasonably productive ponds 2–4 h should be long enough in summer. At the end of the incubation, determine the oxygen concentration in each of the light and dark bottles. Use the following formulae to calculate the amount of inorganic carbon converted to carbohydrates in photosynthesis (carbon fixation) and the amount converted to carbon dioxide in respiration:

$$\text{net productivity} = \frac{(LB - IB)(1000)}{1.2(t)}$$

$$\text{respiration} = \frac{(IB - DB)(1000)}{t}$$

$$\text{gross productivity} = \text{net productivity} + \text{respiration}$$

where productivity and respiration are measured in mg C m^{-3}h^{-1} and IB, LB and DB are the initial oxygen reading, the final (after t hours) light bottle reading and the final dark bottle reading (mg l^{-1}), respectively. An extra dimension can be given to this exercise by incubating several sets of bottles in the same conditions of light and temperature but with different amounts of phosphate nutrient added to each set.

Questions

(1) Why is it advisable to use two floats to suspend the bottles?
(2) What factors might cause the differences between the rates of productivity at each depth in the pond?
(3) Does respiration rate vary with depth? Explain this result.
(4) Pond water almost always contains heterotrophs as well as autotrophs. Explain how their presence in the bottles might affect your results.

Exercise 10: an investigation into the colonisation of surfaces by algae

Background

Whilst some species of microscopic algae are only found suspended in the open water of aquatic habitats, others are adapted for growing on surfaces. Almost every illuminated surface under water carries a living community which, besides the algae, contains bacteria, fungi, protozoa and microscopic metazoans such as nematodes and rotifers.

Many of the organisms in this community attach themselves to the surface on which they are growing and can only be dislodged by scraping or scrubbing. Such forceful techniques destroy the spatial relationships between the organisms and sometimes even the organisms themselves. An alternative approach to studying the attached community is to implant artificial surfaces in the environment, await natural colonisation and then remove the surfaces and take them to the laboratory for further study.

Aim

To study the development of communities on artificial substrates (microscope slides) in fresh water.

Materials

Plain glass microscope slides and coverslips; insulating tape; black polythene; aluminium foil; expanded polystyrene board; string; anchor weight; microscope; centrifuge; tea strainer; 10 % hydrochloric acid.

Time

1 h plus 2 h two weeks later.

Method

Cut pieces from the aluminium foil and polythene sheet using a microscope slide as a pattern. Make twelve sandwiches, each of two numbered slides and a piece of foil or polythene, binding each sandwich together by wrapping insulating tape around the ends. There should be six sandwiches containing each type of filling.

Make three floats from the expanded polystyrene and tie them to the string. Position the floats along the string so that one float will be just beneath the surface, one at mid-depth and one just above the bottom. Leave enough string above the upper float to enable you to retrieve the apparatus from the bank. Attach the anchor to the string. Make four slits in each float and push in four slide sandwiches, arranging it so that one sandwich of each type (foil or polythene) is set horizontally and one vertically (see Fig. 14). Record the positions of the slides.

Gently place the apparatus in a pond and attach the string to something inconspicuous but secure. After two weeks, remove the apparatus from the pond and carefully extract the slides from the floats. A glass staining jar makes a good container for transferring the slides to the laboratory without damaging the attached organisms. Alternatively, slip a small elastic band around each end of each sandwich and lay the slides flat on the bottom of a plastic lunch box. Carry the box back to the laboratory without shaking

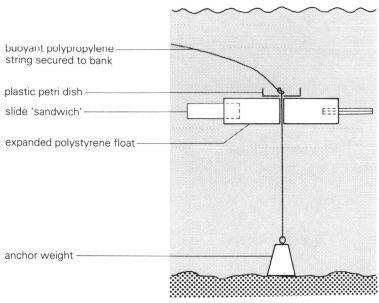

buoyant polypropylene
string secured to bank

plastic petri dish

slide 'sandwich'

expanded polystyrene float

anchor weight

Figure 14   Apparatus for investigating colonisation of microscope slides by algae.

it. Split the sandwiches apart, keeping the outer surfaces of the slides upper-most. Place a coverslip on each slide and examine under the microscope. Identify as many of the organisms as you can. The book by Belcher and Swale (1976; see Bibliography) is useful for identifying algae, but some of the commonest single-celled algae are also shown in Figure 15. Count the number of cells in the field of view or in the squares of a grid, if one is available. Repeat the count for as many fields as necessary to find at least 100 cells and calculate the mean number of cells per field.

The species of diatoms on the slides can be compared with the diatoms of natural attached communities in the same pond by stripping diatoms from pieces of water plant in the following way. Immerse a small piece of plant in a centrifuge tube of hydrochloric acid and leave it in a water bath at 100 °C for ten minutes. Stopper the tube, shake vigorously and separate the resulting suspension of diatoms from the host plant using the tea strainer. Concentrate the diatoms by centrifugation and discard the super-natant. Wash the residue twice by resuspending it in distilled water and cen-trifuging. Remove the supernatant, resuspend the residue in a small volume of water and transfer a drop to a microscope slide. Record the relative abundances of the species of diatom.

This apparatus can also be used to study the development of the attached community over a period of time. In this case, all the slide sandwiches should be of the same type and set in the floats in the same orientation. Sandwiches can then be removed from each float one by one at suitable time intervals.

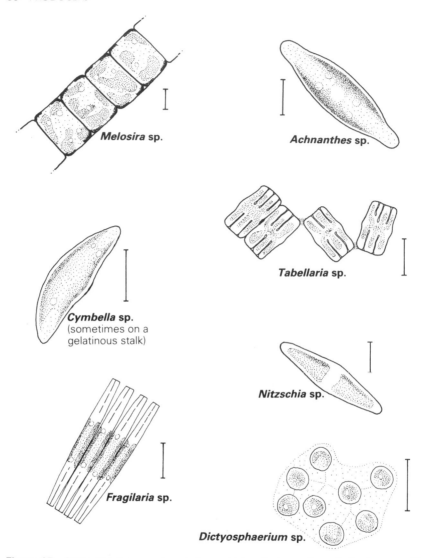

Figure 15 Some common species of algae which colonise surfaces in fresh water. All are diatoms (Bacillariophyceae), except *Dictyosphaerium* which is a green algae (Chlorophyceae). Scale bars represent 20 μm.

Questions

(1) For each type and orientation of slide, explain any differences in the communities which depend on the depth at which they developed.

(2) How do the communities on (a) vertical and horizontal and (b) upward and downward facing slides differ? Explain these results.

(3) Outline and explain any differences in the communities that appear to be due to the material in the centre of the sandwich.

(4) Compare the communities on natural and artificial plants and list the possible disadvantages of using artificial surfaces to study attached communities.

Exercise 11: zonation of macrophytes in ponds

Background

The large plants (macrophytes) at the margins of ponds and lakes often show a pattern of vertical zonation which is reminiscent of the zonation of seaweeds on seashores. On the seashore the pattern is due in large part to the physical conditions found at different heights. These conditions are in turn determined by the twice daily coming and going of the tide. Tides are almost non-existent in fresh water, except in extremely large lakes, so what causes vertical zonation of plants in small lakes and ponds?

Again in common with marine shores, there are horizontal patterns in the distribution of large plants in freshwater environments. These are changes in the plant communities that occur along the length of a shoreline. In this exercise we shall be investigating the nature of both vertical and horizontal zonation.

Aim

To investigate the causes of vertical and horizontal zonation of freshwater macrophytes.

Materials

Weed grapple and depth-measuring device (see Fig. 16); metre rule; large-scale map of pond; white tray.

Time

3 h.

Method

The point at which plants cease to be considered terrestrial and become aquatic is difficult to decide; for our purposes we shall assume that plants that appear to be rooted in water rather than soil are aquatic.

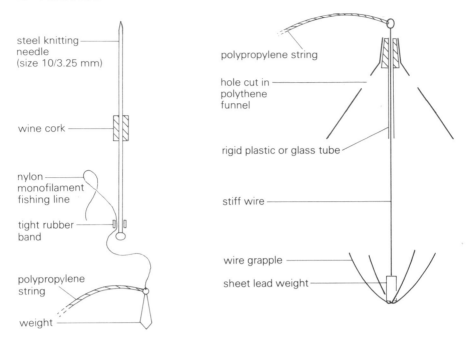

steel knitting needle (size 10/3.25 mm)

polypropylene string

hole cut in polythene funnel

wine cork

rigid plastic or glass tube

nylon monofilament fishing line

stiff wire

tight rubber band

polypropylene string

wire grapple

sheet lead weight

weight

Figure 16 Depth measuring-device and weed grapple. A simpler weed grapple can be made by omitting the guard funnel.

Construct a vertical profile (i.e. a cross section) of the pond along a transect from the water's edge towards the centre. Depths can be measured with the metre rule by wading out as far as possible, or by using the depth-measuring device. To use the depth-measuring device, first guess the depth of water at the first point from which you require a measurement and set the float that depth above the anchor weight. Next, stand at the water's edge and take a firm hold on the rope at the appropriate distance from the weighted end. Now throw the weight out so the rope is stretched out at right angles from the bank. Release some more rope so that the float is directly over the weight. If the float is submerged your original guess was too shallow; adjust the float and try again. If the float lies on its side the guess was too deep, but if the float stands upright your guess was correct.

Find the transition points between different species of macrophyte on the profile. This is straightforward where it is possible to reach the weeds by wading, but in deeper water it is necessary to use the weed grapple. Make the grapple with wire (e.g. a coathanger) that is just stiff enough to catch the weed, but soft enough to bend if it snags something solid. Take hold of the rope at a known distance from the grapple end and throw the grapple out from the bank. The plastic funnel makes sure that the grapple only collects weed from the point where it landed on the bottom. Retrieve the grapple, remove the weed and examine it in the white tray.

Identify the weed using Figure 17 or the key by Haslam *et al.* (1975; see Bibliography). Record whether it was emergent, totally submerged or floating. Mark the distribution of the species on the vertical profile using distribution bars or symbols. Record the bottom type (mud, sand, stones, etc.) as far as you can along the transect and mark it on the profile.

On the large-scale map of the whole pond draw the horizontal zonation of weeds along the shoreline as accurately as you can. If you are working in a large group this information could come from a number of transects around the pond.

Questions

(1) Divide the plant species into emergent, floating-leaved, submerged broad leaved, rosette and submerged linear leaved morphological types, and plot their vertical distribution. Account for their zonation in terms of growing conditions along the transect.

(2) Find out the direction of the prevailing wind and mark it on your map. How might it help explain the horizontal zonation of weeds and bottom types?

(3) Attached algal communities (see Exercise 10) are also found on the leaves of aquatic plants. What effects do you think that these algae might have on the plants?

Figure 17 Identification chart for common pondweeds. Scale bars represent approximately 50 mm. Normal growth habit in brackets; E = emergent, F = floating, S = submerged. ▶

Other common pondweeds include:

Family	Species	Common name and habit	
Ranunculaceae	*Caltha palustris*	marsh marigold	(E)
Nymphaceae	*Nymphaea alba*	white water lily	(F)
	Nuphar lutea	yellow water lily	(F)
Cruciferae	*Rorippa nasturtium-aquaticum*	watercress	(E)
Hippuridaceae	*Hippuris vulgaris*	mare's tail	(S/E)
Umbelliferaceae	*Berula erecta*	water celery	(E/S)
	Oenanthe fistulosa	water drop-wort	(E)
Polygonaceae	*Polygonum amphibium*	amphibious bistort	(F/E)
Boraginaceae	*Myosotis scorpiodes*	water forget-me-not	(S/E)
Scrophulariaceae	*Veronica beccabunga*	brooklime	(E/S)
Lentibulariaceae	*Utricularia vulgaris*	greater bladderwort	(S)
Labiatae	*Mentha aquatica*	water mint	(E)
Alismataceae	*Sagittaria sagittaria*	arrowhead	(S/F/E)
Juncaceae	*Juncus effusus*	soft rush	(E)
Iridaceae	*Iris pseudacorus*	yellow iris	(E)
Cyperaceae	*Eleocharis palustris*	common spikerush	(E)
	Schoenoplectus lacustris	common clubrush	(E)
Gramineae	*Glyceria fluitans*	floating sweet-grass	(F)
Lemnaceae	*Lemna spp.*	duckweed	(F)

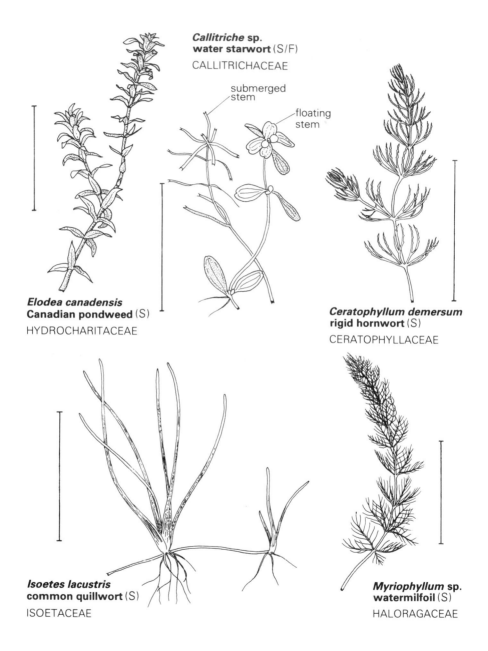

Callitriche sp.
water starwort (S/F)
CALLITRICHACEAE

submerged
stem

floating
stem

Elodea canadensis
Canadian pondweed (S)
HYDROCHARITACEAE

Ceratophyllum demersum
rigid hornwort (S)
CERATOPHYLLACEAE

Isoetes lacustris
common quillwort (S)
ISOETACEAE

Myriophyllum sp.
watermilfoil (S)
HALORAGACEAE

Figure 17

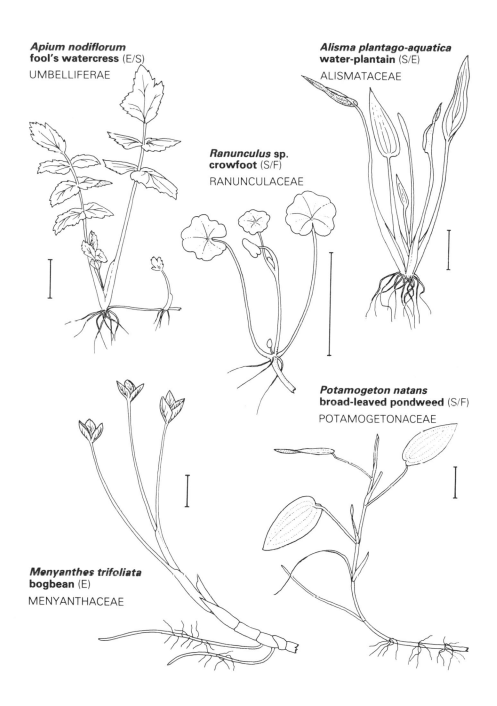

Apium nodiflorum
fool's watercress (E/S)

UMBELLIFERAE

Alisma plantago-aquatica
water-plantain (S/E)

ALISMATACEAE

Ranunculus sp.
crowfoot (S/F)

RANUNCULACEAE

Potamogeton natans
broad-leaved pondweed (S/F)

POTAMOGETONACEAE

Menyanthes trifoliata
bogbean (E)

MENYANTHACEAE

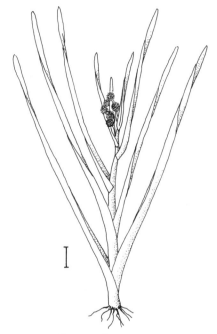

Sparganium erectum
bur-reed (E)
SPARGANIACEAE

Figure 17

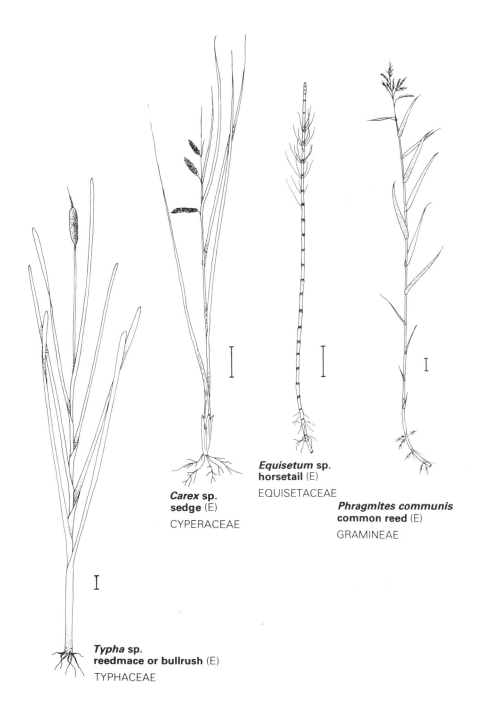

Equisetum sp.
horsetail (E)

EQUISETACEAE

Carex sp.
sedge (E)

CYPERACEAE

Phragmites communis
common reed (E)

GRAMINEAE

Typha sp.
reedmace or bullrush (E)

TYPHACEAE

Consumers

The organisms that feed on other organisms are referred to as consumers. Consumers are often subdivided into a number of groups, each constituting a trophic, or feeding, level. Those that feed on living plants, the herbivores, are called primary consumers, just as the plants are primary producers. Carnivores that feed on herbivores are called secondary consumers, and their predators are tertiary consumers. Primary producers constitute the first trophic level in the system, primary consumers the second, and so on.

Almost every environment, including fresh water, contains consumers that do not fit neatly into this simple scheme of trophic levels. These consumers, the decomposers, feed on the dead remains of other organisms. Without poking into the shadier and less attractive parts of the environment, the vital role of decomposers is easily overlooked, but were it not for their activities we would soon disappear beneath a rising tide of corpses and waste products.

An absence of decomposers would also have a more important but less macabre effect. Plant nutrients, once used to build new tissue, would effectively be locked away and unobtainable to succeeding generations of producers. Decomposers release the nutrients by breaking down the tissue, bridging the gap in the nutrient cycle.

Decomposers are particularly obvious in small, woodland streams. Overhanging trees restrict the light falling on such streams in spring and summer, so primary productivity is reduced and primary consumers are relatively scarce. In autumn, large quantities of dead organic matter enter the stream, both as fallen leaves and as dissolved organic compounds washed in from the surrounding soils. This provides the food supply for abundant populations of decomposers.

Dead leaves are tough and indigestible; dissolved organic compounds are unavailable to larger organisms. Invertebrate decomposers (detritivores), bacteria and fungi work together to process this material in streams.

Freshwater detritivores tend to select their food by particle size rather than by type. They can also be classified by the way in which they feed. Some species, the 'shredders', attack whole leaves, reducing them to finer faecal pellets. Aquatic fungi growing in the leaves are an important part of their diet. The fungi break down the indigestible leaf tissue but are themselves easily digested by the shredders.

'Collector' species gather up the finer fragments, many of which are faecal pellets. All the while, bacteria have been growing on the surface of the pellets, breaking down the indigestible celluloses of leaf tissue. These bacteria are stripped off and digested by the collectors. The material in faecal pellets may pass through this process many times.

Dissolved organic compounds are absorbed by bacteria in the water. These bacteria are caught by the 'filterers', such as the larvae of *Simulium* (see Exercise 5). Their faecal pellets, much larger than the particles they ingest, are gathered and ingested by the collectors. This complex scheme of breakdown of detritus is summarised in Figure 18.

Ponds and lakes also have their shredders breaking down the dead aquatic macrophytes. However, in deep water, where the main source of detritus is dying phytoplankton, the benthic detritivores are mainly collectors and filterers.

In time, the bodies of the decomposers themselves are decomposed. The end products of these breakdown processes are mineral salts and metabolic heat. Mineral salts are taken up as nutrients by growing plants, accumulated in soft sediments, or washed down the rivers to the sea. Metabolic heat is lost as radiant energy to the atmosphere, to be replaced by sunlight trapped by plants in photosynthesis. In some freshwater environments decomposition processes are very slow, so organic matter accumulates in the sediment and eventually forms peat.

As well as food, a suitable habitat must offer a consumer a place in which to live. The spatial distribution of consumers in rivers and lakes often mirrors the distribution of these two resources. In the next exercises we shall investigate the way in which consumers find a new habitat in streams, the use of food by detritivores and the important attributes of a place in which to live.

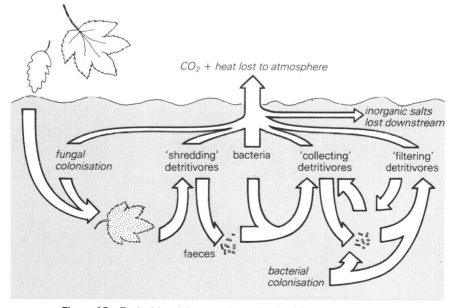

Figure 18 Typical breakdown pathway for dead leaves in streams.

Exercise 12: colonisation of new habitat by stream invertebrates

Background

The invertebrates that live in the streams have two hazards to face. Whilst continuing to feed and to breed, they must avoid being killed by predators or caught by the current and washed downstream. Many choose to live in the spaces, or interstices, between particles of gravel, where they are in almost static water and are less obvious to large, sight-hunting predators such as fish. Some will emerge from the gravel to feed at night, when they cannot be seen. Others find that the interstices serve as a trap for fragments of algae and detritus or as a hunting ground for their animal prey.

If a net is left in a stream so that the current passes through it but the net is not in contact with the bottom, a surprising number of benthic invertebrates will be caught. These animals are being carried along by the current and are said to be 'drifting'. A small proportion may be dead or dying but the majority are usually quite viable. Many, no doubt, lost their footing while feeding and entered the drift unintentionally. The remainder may have intentionally entered the drift, so that they can use the current to carry them to a new feeding area.

Trout feed extensively on drifting animals, which is probably why most invertebrates drift mainly when it is dark. Of course, since many emerge from the gravel to feed at night, it could also be argued that more will be likely to drift by accident then!

Although it is more difficult to demonstrate, it is also known that some stream invertebrates migrate upstream. This movement tends to compensate for the constant shift of the population down stream in the drift. Investigating drift and upstream movement directly requires specialised equipment, but it is possible to study their effects using relatively simple techniques.

Aim

To investigate the colonisation of patches of implanted habitat by stream invertebrates.

Materials

Shallow trays (e.g. plastic seed boxes with the drain holes blocked or screened); buckets; collecting kit (see Appendix 1); wire quadrats (preferably same area as trays); spade; sample jars or polythene bags.

Time

3 h on one day, and an additional ½ h on four occasions over two weeks.

Methods

The principle is to study the rate at which gravel is colonised by collecting the animals that have accumulated in trays which have been in a river for different lengths of time. This can be achieved in either of two ways: the trays can be set out one by one at different intervals before the collection date, or all the trays can be set out at once and individual trays collected on several subsequent dates. In either case, suitable exposure times are 1, 2, 4, 8 and 16 days.

Select an area of river bed which is reasonably uniform and large enough to accommodate all the trays when set about half a metre apart. From a similar site further downstream, dig out enough gravel to fill a tray and place it in the bucket. Fill the bucket two-thirds full with water and swirl the gravel around vigorously. While the water is still swirling, pour it back into the river, retaining the gravel. Repeat the process twice. This removes the resident animals from the gravel.

Dig a shallow depression, large enough to accommodate a tray, at the upstream end of the area of river bed. Fill a tray with the newly washed gravel and set it in the depression. If all the trays are to be set out at once, repeat the gravel washing and tray implantation procedure, but set successive trays downstream of those already in place.

When the time comes to remove the trays (or when the last tray is to be removed, if they were all set out at once), first take samples of the animals from an area of undisturbed stream bed down stream of the trays. Use the quadrat method (see Exercise 6). Next remove the trays, starting from the downstream end and working up stream. As each tray is being removed, hold a pond net behind it so that any animals accidently lost from the tray are caught. Retain the quadrat samples and the tray contents (plus any animals caught in the net) in separate, labelled jars or polythene bags.

Back in the laboratory, pick out, identify and count the animals from the quadrats and trays. Plot the numbers of some of the commoner species in the trays against the time that the trays were in place, to obtain a set of colonisation curves.

This technique can also be used to investigate the effect of many abiotic factors on colonisation and community development. For instance the trays can be provided with different amounts of decaying leaves, filled with gravel of different particle sizes or placed in areas with differing flow conditions.

Questions

(1) Why is it necessary to work in a downstream direction when setting out the trays, but up stream when retrieving them?

(2) Compare the numbers and species of animals from the quadrats and the trays. Account for your results in terms of differences in habitat characteristics.

(3) Which animals are the fastest colonisers? Which are the slowest colonisers? Explain the differences in speed of colonisation.
(4) How might you design an experiment to find whether colonisers walk onto the trays from gravel nearby or travel down from upstream in the drift?

Exercise 13: the fate of dead leaves in streams and ponds

Background

The decomposition of leaves in fresh water can be followed by recording the changes in weight of leaves left immersed for various periods of time. By enclosing the leaves in different sorts of bag, the relative importance of the shredders in the decomposition process can be found.

Aim

To determine the role of shredders in the decomposition of dead leaves.

Materials

Dead leaves, preferably elm or sycamore, collected in autumn and stored dry; a few builder's bricks; nylon or terylene curtain net, approx. 1 and 5 mm mesh sizes (fruiterers' net bags are good for coarse mesh); synthetic string; collecting kit (see Appendix 1).

Time

Small amount of preparation (about 1 h) spread over three months plus 2 h of laboratory work.

Method

Make up 12 small bags (about 20 cm^2) from the net. An office stapler is a convenient substitute for a sewing machine! Select the net so that half the bags have meshes no larger than 1 mm and the remainder have some holes as big as 5 mm. Label the bags.

Dry the leaves in an oven at 40 °C and weigh out batches of about 10 g. Record the weight of each batch and seal it in a bag. Tie the bags together in pairs of one coarse and one fine. At fortnightly intervals tie a pair of bags to a brick in the pond or stream.

At the end of the experiment remove the bags from the water, taking care not to lose any animals or leaf fragments through the meshes. Open each bag and identify and count any animals within. Dry all the animals from each bag and weigh them. Collect the remains of the leaves including any pieces that have become detached. Make a note of the condition of the leaves (intact, shredded, skeletonised, etc.). Dry the remains and weigh them.

Plot the number and the weight (biomass) of the animals in each sort of bag against time. Calculate the weight lost by the leaves in each type of bag from their weights at the beginning and the end of the experiment. Express the weight loss as a percentage of the initial weight. Plot percentage weight lost against time.

Questions

(1) Compare the lists of species from each type of bag. How many in each are shredders (see Fig. 22) and how many are filterers or collectors?
(2) Account for any differences between the plots of number and biomass for each sort of bag.
(3) Explain the changes in weight loss in each type of bag over time.
(4) How does the way in which the leaves decay differ between the fine and coarse mesh bags?

Exercise 14: the food preferences of detritivores

Background

Although detritivores select their food primarily by particle size, they may also choose between different sorts of detritus. Some leaves, like beech, seem to be inherently indigestible, whilst other species are more readily consumed. Since the fungi and bacteria on the leaves seem to be important to the invertebrate consumers, it would not be surprising to find that leaves with a rich microbial flora were preferred.

Aim

To investigate the factors affecting the rate of consumption of dead leaves by invertebrates.

Materials

Scissors; beech, oak, elm, ash and sycamore leaves, collected and stored as in Exercise 13; 100 cm³ crystallising dishes; graph paper.

Time

½ h and 1 h, at least three days apart. A few minutes preparation, two weeks and again, one day in advance.

Method

Soak some dry leaves of each species in pond or stream water for at least two weeks before setting up the experiment. Change the water at least three times. This 'conditioning' process encourages the growth of fungi and bacteria. Soak the remaining leaves for one day to soften them.

Collect detritivores from a pond or stream. The amphipod *Gammarus* or the isopod *Asellus* are good subjects for this experiment.

Cut 20 mm squares from conditioned and unconditioned leaves of each species, using a graph paper template. Set up two dishes for each species of leaf, one for conditioned squares and the other for other unconditioned squares. Place three squares and six *Gammarus* or *Asellus* in each dish. Keep the dishes in a cool, dark place.

After three days, use the graph paper template to measure the area of each leaf square that has been eaten. Calculate the total leaf area consumed in each dish. Plot the results as a bar graph.

Questions

(1) Is there any difference between the rates at which each leaf species is consumed? Discuss the factors which might cause these differences.
(2) What effect has the conditioning process had on the rate of consumption? Are all the leaves affected in the same way?
(3) How would you design an experiment to find which leaves were preferred by the detritivores when they are given a choice? What would you expect the results to be?
(4) What other factors may have influenced the amount of leaf consumed in each dish? How could the experimental design be improved?
(5) Discuss the advantages and disadvantages of recording consumption by measuring area rather than by weighing, as in Exercise 13.

Exercise 15: weed beds as habitats for invertebrates

Background

Weed beds often carry dense populations of many species of invertebrate. The weeds may be valued as a habitat for two reasons. First, the herbivores may actually be grazing on the weeds, so that they are attracted by the abundant source of food. The predators, of course, follow the herbivores for the same reason. Second, the weeds may merely be providing a place in which to live, just like the gravel in the river bed. Here it is the physical characteristics of the weed that are important.

We can attempt to find out which of these two roles the weed is playing by offering invertebrates artificial weeds. These provide similar physical characteristics to the real thing but are not suitable as food.

Aim

To determine the important characteristics of aquatic plants as habitats for invertebrates.

Materials

Polypropylene string; a wooden or metal stake; microscope, slides and coverslips; collecting kit (see Appendix 1).

Time

2 h plus 1 h preparation two weeks in advance.

Methods

Find an area of a stream or pond with beds of submerged weeds. By knotting together lengths of string, make a reasonable imitation of the weed. Drive in the stake and attach the artificial weed.

Remove the artificial weed after two weeks. If in a stream, do this by sweeping it up in the net from the downstream end and cutting it at the stake. Collect samples of a similar bulk of real weed in the same way. It may first be necessary to tease out a skein from a larger bed.

Examine the real and artificial weeds leaf by leaf and remove all the animals you can find. Retain the weeds. Identify, count and list the animals from each sort of weed. Gently scrape any material adhering to the surface of the artificial weed onto a microscope slide and examine under medium power. Make a note of what you see.

A variation on this exercise is to make artificial weeds of different sizes and shapes, in order to investigate which aspects of the plant architecture are particularly important to colonising invertebrates. It is also possible to use artificial weeds to study colonisation processes (cf. Exercise 12).

Questions

(1) From your results, suggest which is the more important role of the weed, as food or as a place in which to live.
(2) List the organisms scraped from the surface of the artificial weed. What do you think might be the food source for the consumers found here?

Interactions and community structure

The group of populations living together in a particular place is called an ecological community. An isolated population with abundant food would grow at a rate governed by its intrinsic rate of reproduction and its rate of death through old age. In communities, interactions tend to change natural lifespans and reproductive rates.

When resources for a population are in short supply, its members may compete between themselves for dwindling stocks. This interaction, called intraspecific competition, would eventually restrict population growth and set the upper limit on the size of an isolated population. In a community, competition for a limiting resource can also occur between populations of two or more different species. This is known as interspecific competition.

Predation is an interaction which enhances the growth and reproduction of the predator but decreases the lifespan of the prey. Herbivores can be considered as predators of plants, but often they cause injury rather than death. The relationship between a parasite and its host is also similar, but a parasite is dependent on its host to provide the environment in which it lives and must take care not to cause its premature demise. The prudent parasite exacts a living with the minimum disruption to the population processes of its host.

Occasionally, the interaction between two species is such that neither party is detrimentally affected. If the interaction is to the benefit of one species it is called commensalism. When both parties benefit from the interaction it is called mutualism. Mutualistic interactions may eventually reach very intimate levels of involvement, so that the two species cannot exist alone.

The costs or benefits of all these interactions to each of the species are summarised in Figure 19.

Often the species list and the relative abundance of the populations in a community are remarkably constant. These are aspects of the structure of the community. The species list is governed, in part, by tolerances for the abiotic conditions offered by the environment. Interactions between those species that are capable of tolerating the local physical and chemical conditions provide a second, biotic, influence on the structure of communities.

| Interaction type | Effect on | |
	Species A	Species B
mutualism	+	+
commensalism	+	0
parasitism	+	−
predation	+	−
competition	−	−

Figure 19 Types of ecological interactions between pairs of species. A plus sign indicates that the interaction has a beneficial effect on one of the species, a minus sign a detrimental effect.

Exercise 16: a study of predation by damselfly larvae

Background

Brightly-coloured adult damselflies are a familiar summer sight, flashing along the banks of rivers and ponds. For the rest of the year, damselflies are to be found as rather drab and ponderous larvae, which stalk the mud and weed below the water's edge.

Whether as adults or larvae, damselflies are formidable predators. Like their larger cousins, the dragonflies, they are equipped with a special extensible jaw called the mask. This is a modification of the lower mouthpart, armed with a pair of pincers on the end, and is normally kept folded beneath the head. When a suitable prey passes within range, the mask is shot out at great speed and the prey is impaled on the pincers (see Fig. 20).

Aim

To study the rate at which larval damselflies consume Cladocera under laboratory conditions.

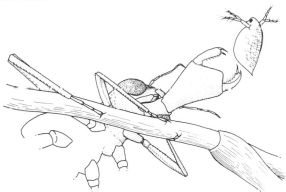

Figure 20 Damselfly nymph catching a cladoceran. Note the specially adapted labium or 'mask'

Materials

Containers, such as beakers, about 100 cm³ capacity; twigs or cocktail sticks, weighted to sink; Pasteur pipettes; collecting kit (see Appendix 1).

Time

1 h on each of two occasions 24 h apart.

Method

Damselfly larvae are usually commonest in weedy ponds or in weed beds in slow-flowing rivers. Collect enough to provide one per container. The same site will probably provide *Daphnia* or a related cladoceran, such as *Bosmina*. A fine-mesh (<250 μm) pond net, swept in and around the weeds, should catch an ample supply.

Fill each container with 50 cm³ water and add a twig or cocktail stick to act as a perch. Gently place one larva in each container and leave it for an hour to recover. Use the pipette to add a different, known number of *Daphnia* to each container. Try densities of 2, 5, 10, 20 and 50 and 100 per container. Make up the volume to 75 cm³ with water.

Leave the containers in a cool, dark place for 24 hours. Count the number of *Daphnia* remaining after this time and obtain the number eaten by subtraction. Plot the number consumed per hour against the density of prey.

Questions

(1) How does the rate of prey capture change with increasing density of prey? List the factors that may be causing this result.
(2) How do the damselflies detect their prey? Devise an experiment to test your hypothesis.
(3) Observe what happens to the distribution of *Daphnia* if they are illuminated from one side. Use this observation to explain why the experiment was conducted in the dark.

Exercise 17: how a mollusc avoids its predators

Background

The common bladder snail, *Physa fontinalis*, falls prey to a variety of invertebrate predators. In order to reduce the risk of death it has evolved a special anti-predator defence. When a predator is detected, *Physa* shakes its shell vigorously to and fro and may eventually detach itself from the surface on which it was moving. Once detached the snail will either float to the surface of the pond or else sink quickly to the bottom, depending on how much air is held in the mantle cavity within the shell. In either case the snail effects escape from its prospective predator.

Aim

To study the response of *Physa fontinalis* to prospective predatory invertebrates.

Time

1 h.

Materials

9 cm diameter Petri dishes; glass stirring rod; collecting kit (see Appendix 1).

Method

Collect and identify about 50 bladder snails (see Key 3). Weedy ponds or streams with clear, unpolluted water are good places to look. The snails can be returned unharmed at the end of the exercise. Also collect and identify (see Key 1) a few representatives of as many leech and flatworm species as possible. Leeches and flatworms will often be found attached to the bottom of an enamel or plastic sorting tray after a sample of weed or mud has been poured away.

Testing for the predator avoidance response requires one snail in each Petri dish of pond water. Allow the snails to settle for at least 30 min. Gently touch the exposed lobes of the mantle tissue, which are normally wrapped around the last whorl of the shell, with part of the prospective predator. The anterior part of a leech can be used or the whole animal in the case of a flatworm. Observe the response of the snail and score it as strong if the snail detaches from the dish and weak if only gentle shell shaking is evoked. If nothing happens within 30 seconds of touching the mantle tissue, score it as a null response. Use each snail only once and arrange the experiment so that each prospective predator is tested the same number of times.

Plot the percentage of each type of response as shaded columns in a bar graph or sectors in a pie chart for each predator. Test for differences between the percentages of strong and weak or null responses using a chi-square test on a 2×2 contingency table. For example:

Response	Predator 1	Predator 2	Total
strong	60(A)	30(B)	90
weak or null	40(C)	70(D)	110
total	100	100	200(N)

$$\chi^2 = \frac{N([AD - BC] - N/2)^2}{(A + B)(C + D)(A + C)(B + D)}$$

$$= \frac{200([4200 - 1200] - 100)^2}{(60 + 30)(40 + 70)(60 + 40)(30 + 70)}$$

$$= 16.99$$

The degrees of freedom for this test are given by the number of rows in the table minus one, multiplied by the number of columns minus one. In this case the experimental value of chi-square is greater than 3.84, which is the value for one degree of freedom and a probability of 5 % (i.e. $p = 0.05$) taken from Figure 21. This means that if there was in reality no difference between the responses of the snail to the two predators (the null hypothesis) a difference of the sort observed in this experiment would be likely to occur by chance alone in less than 5 % of all experiments. In fact, the observed value of chi-square is greater than the tabulated value at a probability of 0.1 % (10.83), so the observed result could be expected to occur less than one in a thousand times if the null hypothesis were true. Therefore it is reasonable to reject the null hypothesis and conclude that in reality the snail's response to each predator is different.

If one species of predator consistently evokes a strong response, devise experiments to find which parts of the snail are receptive to the stimulus and which parts of the predator are capable of triggering a response. Mucus from the predator can be tested by first transferring some to a glass stirring rod.

Questions

(1) Compare the results of the predator tests with the normal diets of the predators (see Fig. 22). What do you conclude?
(2) Discuss the disadvantages to *Physa* of an unspecific predator response.
(3) Test the response of *Physa* to other individuals of its own species. What might be the consequences of the behaviour it displays, in terms of the distribution of individuals in the habitat? How might this be an adaptive response to predation by fish, predators against which shell shaking is ineffective?

p	0.90	0.50	0.10	0.05	0.01	0.001
Chi-square	0.016	0.46	2.71	3.84	6.64	10.83

Figure 21 Values of chi-square and their probabilities for one degree of freedom.

Animal type	Feeding method	Foods
sponge	F	protozoa, small organic particles
Hydra	E	Cladocera, copepods
rotifers	F	protozoa, small organic particles
Oligochaetes	C	fine organic detritus
Hirudinea (leeches) Hirudinidae, Erpobdellidae	E	chironomid larvae, molluscs, oligochaetes
Piscicola, Hemiclepsis	B	fish,' Amphibia
Helobdella, Glossiphonia	B	chironomid larvae, oligochaetes, amphipods, gastropods
Theromyzon	B	ducks
nematodes	C	fine detritus
gastropods	SC	benthic algae, detritus
bivalves	F	planktonic algae
Coleoptera (beetles) Hydrophilidae	C	algae, detritus
Gyrinidae	E	small invertebrates
Elminthidae	SC	benthic algae, detritus
Haliplidae	SH	macrophytes, algae
Dytiscidae	P	invertebrates, small fish, tadpoles
Hemiptera (bugs) water measurer, water cricket, pond skater, *Notonecta*	P	Small aquatic and terrestrial invertebrates
water scorpion, water stick insect	P	small aquatic invertebrates
Corixidae	C	algae, fine detritus
Ostracoda	C	small detritus particles
Cladocera	F	planktonic algae, detritus
Copepoda	E	algae, small invertebrates
Asellus, Gammarus	SH	large detritus particles
crayfish	SH	decaying organic material

Figure 22 Typical feeding methods and foods of some freshwater invertebrates. Feeding methods are as follows: B, blood-sucking; C, collecting from sediment surface; E, engulfing, grabbing and swallowing prey; F, filtering; P, piercing prey and sucking out predigested body contents; SC, scraping organic matter from hard surfaces; SH, shredding large particles of organic detritus, often dead tree leaves.

Figure 22 continued

Animal type	Feeding method	Foods
Insect larvae		
Chaoborus	E	Cladocera, copepods
Culicidae	F	algae, protozoa, detritus
Tipulidae	SH, C	large detritus particles
Simulium	F	bacteria, fine detritus
Chironomidae		
predatory	E	insect larvae, oligochaetes
non-predatory	C	fine detritus, algae
Plecoptera (stoneflies)		
predatory	E	smaller insect larvae
non-predatory	C, SH, SC	detritus, algae
Odonata	E	small crustacea, insect larvae
Ephemeroptera (mayflies)	SC, C	algae, fine detritus
Trichoptera (caddisflies)		
Polycentropodidae	F	small invertebrates
Rhyacophilidae	E	mayfly, blackfly and midge larvae
Hydropsychidae	F	small invertebrates, detritus
Hydroptilidae,		
Glossosomatidae	SC	algae, fine detritus
other cased caddis	SH, C	mainly detritus
Coleoptera (beetles)		
Haliplidae	SH	macrophytes, algae
Helodidae	SC	algae
Helminthidae	SC, C	algae, fine detritus

Exercise 18: competition between species of Cladocera

Background

The effect of intraspecific competition on the growth of populations of *Daphnia* was investigated in Exercise 8. The same techniques can be used to culture mixed populations of two species of Cladocera. If desired, this exercise could be carried out as an extension of Exercise 8.

Materials and time

See Exercise 8.

Method

Use the culture methods described in Exercise 8. Collect sufficient individuals of two species of Cladocera from a pond, or alternatively culture them separately in the laboratory and remove surplus animals for use in the mixed experiments.

All culture flasks should contain the same volume of medium and the same concentration of yeast. One-third of the culture flasks should be started with five individuals of one species and one-third of the flasks with five of the second species. The remaining flasks should be started with five individuals from each of the two species.

At intervals of three to five days transfer the occupants of each flask to a freshly prepared flask, counting and identifying them as you do so. Repeat this procedure until the populations in the flasks have reached an apparently stable level (which may be extinction). Plot the mean population size, including zero values, against time using a separate piece of graph paper for each species. Each graph should have two lines, one for the population size when the species was cultured alone and the other for the result of the mixed species culture.

Questions

(1) What evidence is there for interspecific competition in this experiment?
(2) Which resource (food, space, etc.) do you think the species are competing for? Design an experiment to test your hypothesis.
(3) If one or other of the species became extinct in the mixed culture, how do you explain the co-existence of these two in nature?

Exercise 19: the parasites of freshwater shrimps

Background

Thanks to modern standards of hygiene and medical care, western man is often unaware of the prevalence of parasites in natural populations. Most animals and plants are the unwilling hosts of large numbers of species of parasites. In some cases these parasites attach to the external surface of the host, and are called ectoparasites, whilst in others the parasites live within the host's tissues and are called endoparasites. The freshwater shrimp *Gammarus pulex* can host a wide range of ectoparasites and endoparasites. Several species of ciliate protozoans attach to the cuticle of *Gammarus*, and three species of parasitic worm of the Phylum Acanthocephala (meaning spiny-headed) reside within the body cavity. All of these are relatively easy to see.

Aim

To study the ectoparasites and endoparasites of the freshwater shrimp *Gammarus pulex*.

Materials

Collecting kit (see Appendix 1); microscope, slides and coverslips; soda siphon or source of carbon dioxide; saline solution; sodium tauroglyco- cholate (optional).

Time 1 h.

Methods

Collect several hundred *Gammarus* from a stream, preferably near to a duck pond.

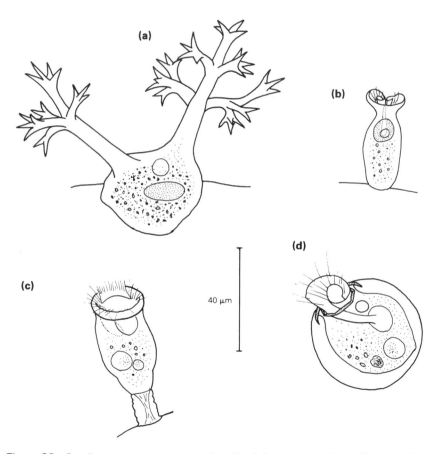

Figure 23 Sessile protozoa common on the gills of *Gammarus pulex* (rotifers may also be present). All are ciliates, although *Dendrocometes* only carries cilia in its larval stage: (a) *Dendrocometes* sp. (Order Suctorida), a predator which catches motile ciliates on the sticky ends of its tentacles and sucks out their cell contents; (b) *Spirochona* sp. (Order Chonotrichida), a filter feeder on bacteria; (c) *Epistylis* sp. (Order Peritrichida), a stalked species which closes the anterior filter feeding apparatus when disturbed; (d) *Lagenophrys* sp. (Order Peritrichida), lives in a hemispherical case with a purse-like opening through which the ciliated filter feeding apparatus can be withdrawn. In each drawing the uppermost, un- shaded, vacuole is the contractile vacuole.

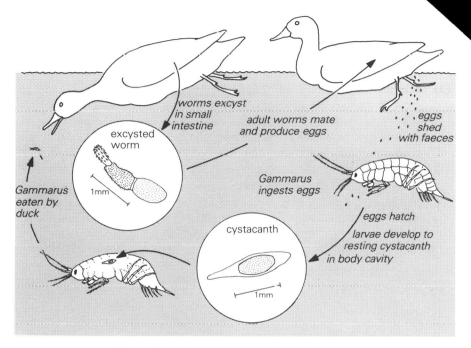

worms excyst
in small
intestine

adult worms mate
and produce eggs

eggs
shed
with faeces

excysted
worm

1mm

Gammarus
eaten by
duck

Gammarus
ingests eggs

eggs hatch

larvae develop to
resting cystacanth
in body cavity

cystacanth

1mm

Figure 24   The life cycle of *Polymorphus minutus*.

In the laboratory, anaesthetise the animals by leaving them in carbonated water from the soda siphon. The same effect can be obtained by bubbling carbon dioxide through water. Examine the surface of the shrimps for attached protozoans, either by viewing directly if a suitable binocular dissection microscope is available, or by removing parts and mounting them on slides for examination under a transmitted light microscope. Record the distribution of protozoans and the number of types present (some common species are shown in Fig. 23).

Acanthocephalan endoparasites can usually be seen through the body wall of the shrimps, if they are present. Fish are the final hosts of the species *Echinorhynchus truttae* and *Pomphorhynchus laevis* and their intermediate stages in the shrimps are white in colour. The final hosts of *Polymorphus minutus* are ducks and this species is easily recognised in the shrimps by its orange colour. *Polymorphus* is known to cause sterility in female *Gammarus* and in this way may affect the population dynamics of the shrimp. Figure 24 shows the life history of *Polymorphus minutus*. An article by Moore (1984; see Bibliography) provides further information on acanthocephalan parasites.

Dissect out cystacanths of *Polymorphus minutus* from infected shrimps. If the cystacanths are kept in warm saline (38 °C), excystment and eversion of the spiny proboscis will often take place. Sometimes this process can be stimulated by keeping the cystacanths in a 0.05 % solution of sodium tauroglycocholate in buffered saline.

pattern of distribution of protozoans on the shrimps.
the protozoa be harming the shrimps (they do not feed on
s' tissues)?

hropods, *Gammarus* has to shed its exoskeleton in order to
ze. What problems does this cause for ectoparasites?

(4) Why is it necessary to warm the cystacanths or expose them to sodium-tauroglycocholate (a vertebrate bile salt) in order to stimulate the final stage of development of *Polymorphus*?

(5) The cystacanth stage of some species of *Polymorphus* is known to affect the behaviour of the host shrimp. Instead of being negatively phototactic and staying near the bottom of the pond or stream, infected shrimps swim near the surface more often than uninfected shrimps. Suggest how this might benefit the parasite.

Exercise 20: community structure

Background

Two aspects of the structure of an ecological community, namely its species composition and their relative abundances, have already been mentioned. Community structure can also be described in the form of a food chain and web or by analysing the community in terms of its trophic levels (see p. 00).

Food chains and webs trace the feeding relationships between the species in the community. A food chain may originate with either a living plant or a source of organic detritus and continue through a herbivore or detritivore until the last link of the chain, the top predator, is reached. It is unusual to find a food chain with more than five links. The reason for this is not fully understood but, since only a small proportion of the food energy is passed on through each link, at some point there may not be sufficient food to support another predator population. Communities rarely contain a single food chain, instead many chains are interwoven to form a food web. An example of a freshwater food web is shown in Figure 25.

If all the species in a community are grouped together in their trophic levels, the structure of the community can be viewed in terms of the relative importance of each trophic level. The presence of detritivores can complicate this analysis, but if the detritus originated from plant growth within the community it is reasonable to class detritivores as primary consumers.

The importance of a trophic level can be measured in a number of ways, the most useful of which is the amount of energy that passes through it. Energy flow is a valuable measure because it is a common currency that can be used to compare communities that differ greatly in their species. Furthermore, energy is not recycled within communities. When the energy flow

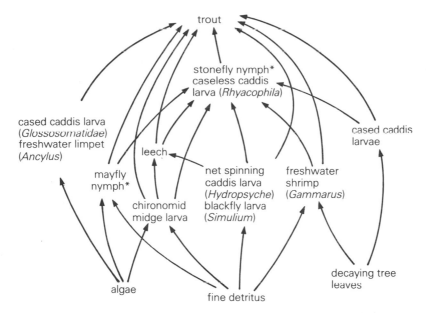

Figure 25 The food web of a stony stream. Stages marked with asterisk are strictly immature stages of insect which do not have a pupal stage in their development (e.g. Plecoptera, Ephemeroptera, Odonata) and are called nymphs rather than larvae.

through the trophic levels is visualised in a bar graph, a characteristic pyramid shape is seen, an ecological pyramid of energy (see Fig. 26).

Alternatively, the importance of each trophic level can be measured in terms of the numbers of individuals or the biomass that it contains. Although these measures can also yield ecological pyramids (of numbers or biomass) the properties of the community are sometimes such that this pyramid structure is lost. Pyramids of numbers are affected when a large number of small consumers is supported by a few large producers (e.g. greenfly on roses), and pyramids of biomass become distorted when a small biomass of producers grows and reproduces fast enough to support a larger biomass of consumers (e.g. phytoplankton and zooplankton).

Although the trophic level approach provides an overview of the number, biomass or energy relations of the community, it takes no account of the ecology of individual species. Some aspects of the ecology of the species can affect trophic relations within the community in ways that cannot be predicted from a knowledge of the shapes of ecological pyramids. Important aspects of community structure and function are concealed within the trophic level divisions and this rather limits the appeal of the approach.

Aims

To investigate the structure of a weed bed community in a pond.

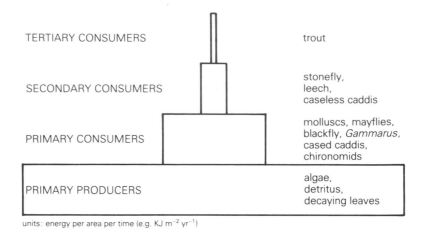

units: energy per area per time (e.g. KJ m⁻² yr⁻¹)

Figure 26 A pyramid of energy for a stony stream community

Materials

List of species and their abundances from a weed bed community (suitable data can be collected as part of Exercise 1); small scalpel or mounted needle; microscope, slides and coverslips; oven; balance.

Time

1 h if data are at hand.

Method

Use the table of foods and feeding methods to establish the feeding relationships between the species in the community (see Fig. 22). Remember that a species may have more than one food source. Display the results in a food web diagram, arranging it so that as few of the lines cross as possible.

Expected feeding relationships can sometimes be confirmed by gut analysis. Use a small scalpel or mounted needle to dissect out the gut. Lay the gut intact on a microscope slide. Carefully open the gut, add a drop of water and scrape the gut contents into it. Put a coverslip on the slide and examine under a microscope. Try to identify the contents.

Group the organisms into trophic levels. Add up the numbers of organisms in each trophic level, if necessary making a 'guesstimate'. Estimate the biomass of each species of plant and animal by multiplying the average dry weight of an individual by its population size or density. Dry the organisms at 105 °C until the weight is constant. Very small organisms should be dried and weighed in batches containing known numbers. Species biomasses can then be added together to give the biomass of the trophic level. Plot the numbers and biomasses as bar graphs (cf. Fig. 26).

Questions

(1) How does population density relate to the species' position in the community food web? Explain your observations.

(2) Why might gut contents give a false impression of a diet? In which sorts of animals would visual examination of gut contents reveal nothing at all about their diet?

(3) Compare the shapes of the pyramids of numbers and biomass and explain any differences.

(4) List any important members of the community that your sampling method is likely to have missed. How might their inclusion in the results change the appearance of the food web or the pyramids of numbers, biomass and energy?

Pollution and waste treatment

In Britain, water is used at the rate of about 300 l per person every day. Of course not all of this is used in domestic consumption! Most is used by industry in servicing the needs of a population with a Western European standard of living.

The annual consumption of water is about one-tenth of the annual residual rainfall in the British Isles. Residual rainfall is the difference between total precipitation and the amount that is 'lost' by evaporation or through transpiration by plants – in other words, the water that is available for use by man. Unfortunately, most of our residual rainfall occurs in places where the demand for water is not very great (mainly in the upland areas of western Britain) and at times of the year when usage is low (in winter). So in the drier and more populous parts of Britain, demand for water often exceeds the local supply. In many places the demand can only be satisfied by using the same water over and over again before it finally reaches the sea.

Almost any use of water influences its quality. Used water is generally returned to a river for disposal. It may have undesirable effects on the life in that river if it differs in quality from the river water. In a country in which such a large proportion of available water is used, it is essential that the effluent is clean. Quite apart from the necessity to protect river life, a high quality effluent is required if the water is to be re-used, perhaps for drinking, somewhere further down stream.

Pollution is not an easy concept to define. For our purposes pollution may be said to occur when man causes a change in a freshwater environment which has undesirable consequences, either to the living systems within it or to those who make reasonable and lawful use of that environment. In this section we shall consider some of the common causes and effects of freshwater pollution and the steps that can be taken to minimise the polluting effects of effluents.

Exercise 21: measuring the quality of an effluent

Background

Many of the uses to which water is put cause an increase in the amount of dissolved or suspended organic matter. This is particularly true of domestic uses, but it also applies to industrial processes which involve living materials, such as abattoirs, dairies and paper mills.

Organic materials form a rich food supply for bacteria which are naturally present in fresh water. Respiration by the bacteria can cause dramatic reductions in the amount of oxygen dissolved in the water. Other organisms can be adversely affected, not only by the shortage of oxygen, but also by the blanketing or clogging effects of large numbers of bacteria.

A common method of assessing the extent of organic pollution is to measure the amount of oxygen consumed when the organic matter in a water sample is oxidised. When a sample is stored for a period of time oxidation occurs naturally, partly through chemical processes but mainly through respiration by bacteria. The amount of oxygen consumed in this way is called the biochemical oxygen demand (BOD).

Aim

To assess the quality of an effluent by finding its biochemical oxygen demand.

Materials

250 cm³ reagent bottles with ground glass stoppers; oxygen meter or Winkler titration reagents and apparatus (see Appendix 2).

Time

1 h for collection and preparation of samples, 1 h for analysis about five days later (less if an oxygen meter is available).

Method

Organically polluted effluents are produced by sewage works, farms and food processing plants. If a source of organic pollution on a local river cannot be found, then an 'artificial' effluent can be used, e.g. used water from school kitchens' sinks.

Safety note: take precautions to avoid contact with water polluted with sewage; wear rubber or polythene gloves and wash hands thoroughly. Disinfect or sterilise contaminated glassware.

Fill the reagent bottles to the brim with samples of polluted and unpolluted water. Two bottles are required for each water source. Insert the lightly greased stoppers so as not to trap air bubbles in the bottles. If a polluted river is being studied, the samples should come from sites on the river up stream and down stream of the effluent source. Measure the temperature of the water from each site.

Fix the dissolved oxygen in one of the bottles from each source using steps 1 and 2 of the Winkler technique described in Appendix 2. The fixed bottles can either be stored together with the incubated bottles, and the remainder of the Winkler technique carried out a week later, or the oxygen content of the fixed bottles can be determined straight away. Alternatively, use an oxygen meter to measure the dissolved oxygen in one of the bottles from each pair.

Incubate the remaining bottles at about 20 °C in the dark. The conditions inside a light-tight cardboard box at normal room temperature are quite satisfactory. Store the bottles under water in a bucket or basin so as to prevent the entry of atmospheric oxygen. The standard incubation period of five days is deemed to be the minimum necessary for more or less complete oxidation of organic matter, but no harm will be done if the samples are incubated for longer.

After incubation, fix the oxygen in the untreated bottles and proceed with the Winkler analysis. If using an oxygen meter, measure the dissolved oxygen content of the incubated samples. Calculate the BOD by subtracting the final oxygen concentration after incubation from the initial value.

Questions

(1) Why is it necessary to incubate the samples in the dark?

(2) If the method described above is used, the maximum BOD that can be measured is about $15 \text{ mg} l^{-1}$. Explain this and suggest a modification of the method that would allow measurement of a higher BOD.

(3) Some experts recommend that all samples should be 'seeded' with bacteria before incubation. Why might this be necessary? Criticise the use of the BOD technique to measure the impact of an effluent on oxygen concentrations in a river.

(4) Oxygen depletion due to organic pollution is usually most severe in summer. Suggest three reasons for this.

(5) A chemical oxidation method can be used instead of relying on oxidation of organic matter by bacteria. The oxygen consumed is then called the chemical oxygen demand (COD). Why might this method give a false indication of the potential impact of the effluent on oxygen conditions in a river?

Exercise 22: the effect of pollution on a river community

Background

Although it is technically possible to measure the concentration of almost any pollutant in river water, the information obtained does not always give a useful indication of the impact of the pollution on river life. This is so for two reasons. First, water samples are taken at one point in time and may not fairly represent the worst, or even the average, conditions to which the river community is exposed. Secondly, the effect of one pollutant on an organism is often modified by the presence of other pollutants. This is called a synergism between pollutants.

A better indication of the conditions in a river is often obtained by studying the river community. The plants and animals that live in the river 'sample' the water continuously and their presence or absence can be used as a measure of river purity. In order to make the information that comes from a study of river communities more digestible, particularly for non-biologists with an interest in river management, many indexes of river pollution have been devised. These combine the presence or absence of many species, and sometimes their abundances, in one single index value. In this exercise we shall use the Trent biotic index, which is based solely on presence or absence of certain groups of animals.

Aim

To assess the status of a river community with respect to organic pollution, using the Trent biotic index.

Materials

Collecting kit (see Appendix 1); pH meter or test papers; thermometer.

Time

One day.

Method

Choose a reach of a river in which there is a recognisable potential source of pollution. Collect 'kick' or 'sweep' samples of the bottom-dwelling invertebrates from at least one site upstream of the source of pollution and from a series of sites at different distances down stream. One of the downstream sites should be close to the effluent; one should be as far down stream as it is practicable to go (as long as the general character of the river remains constant); the remainder should be spread at roughly equal intervals in between. Try to collect from the same sort of bottom at each site.

At each site measure the water temperature and pH. If water samples are collected this exercise can be combined with Exercise 21.

In the laboratory, sort the samples and identify the animals. The basic units of the Trent biotic index are the groups defined at the bottom of

Figure 27. Note that it is usually not necessary to name species, but that it is important to know how many species are present. Generally, two individuals that clearly differ in some characteristic other than size can safely be assumed to belong to separate species. *Nais* worms are mostly less than 2 cm long and transparent, whereas tubificid worms are usually longer and red in colour. If there is only one Baetid mayfly species it is most likely to be *Baetis rhodani*.

Trent biotic index values are calculated by proceeding down column 1 until an appropriate line is reached, e.g. if there were no stoneflies or mayflies, but there were caddis-flies, then either line 5 or 6 would be correct. The choice between line 5 or 6 in this case would be made by referring to the categories in column 2. The biotic index value of the community is found in columns 3 to 7, depending on the total number of groups present. Note that groups named at the top of column 1 are least tolerant of organic pollution, while pollution-tolerant groups are found towards the bottom of the column.

Figure 28 shows comparative values of the Trent biotic index, BOD, community composition and overall pollution assessment which are drawn from experience of river monitoring. Display the results of this exercise as bar diagrams of species distribution along this reach of the river. Plot

Line no.	Column no.: 1 Indicator taxa present	2 Diversity of fauna	3	4	5	6	7
					Total no. of groups		
			0–1	2–5	6–10	11–15	16+
					Biotic index		
1	Plecoptera nymphs	more than one species	—	VII	VIII	IX	X
2		only one species	—	VI	VII	VII	IX
3	Ephemeroptera nymphs	more than one species	—	VI	VII	VIII	IX
4		only one species	—	V	VI	VII	VIII
5	Trichoptera larvae	more than one species	—	V	VI	VII	VIII
6		only one species	IV	IV	V	VI	VII
7	*Gammarus*	above taxa absent	III	IV	V	VI	VII
8	*Asellus*	above taxa absent	II	III	IV	V	VI
9	tubificid worms and/or red chironomid larvae	above taxa absent	I	II	III	IV	—
10	above taxa absent	some organisms not needing dissolved O_2 may be present	0	I	II	—	—

Figure 27 Table for calculating the Trent Biotic Index. Note that *Baetis rhodani* (Ephemeroptera) is more tolerant of organic pollution than the other mayfly species and is included with the Trichoptera in columns 1 & 2. The following taxa constitute groups when calculating the Trent Biotic Index:

each species of Platyhelminthes, Hirudinea, Mollusca, Crustacea, Plecoptera, Neuroptera, Coleoptera, Hydracarina and Diptera (except Chironomidae and Simuliidae);
each genus of Ephemeroptera (except *Baetis rhodani*);
each family of Trichoptera;
Family Chironomidae (except red chironomids), red chironomids,
Family Simuliidae, Class Oligochaeta (except genus *Nais*), *Nais*, *Baetis rhodani*.

Trent index	BOD (mg l^{-1})	Condition	Characteristic fauna
XI–X	2	very clean	trout & salmon, Plecoptera, Ephemeroptera, Trichoptera, Amphipoda
VII–X	2–3	clean	coarse fish, plus the groups above
VI–VIII	2–3	clean	coarse fish, but fewer species of the groups above except for Trichoptera, Amphipoda and Baetid mayflies
V–VI	3–5	fairly clean	few fish: Trichoptera restricted; baetids rare – *Asellus* dominates, with Mollusca and Hirudinea present
III–V	5–10	doubtful	as above, but with no baetids
II–IV	5–10	doubtful	as above, but no fish
I–III	10 +	bad	only red chironomids and Oligochaeta
0–I	10 +	bad	only species capable of breathing atmospheric oxygen

Figure 28 Comparative values of the Trent Biotic Index and other indicators of pollution.

graphs of temperature and pH (and BOD if measured) against distance down stream.

Questions

(1) Discuss the status of this river with respect to pollution.
(2) The Trent biotic index, like most others, is designed to assess the severity of organic pollution. In what circumstances might it give misleading results if applied uncritically?
(3) Why is it necessary to sample from the same type of bottom at each site?
(4) How might natural changes in river characteristics complicate the investigation of the effects of pollution? List any factors that may be obscuring your investigation of pollution.

Exercise 23: an investigation of filter bed communities in sewage works

Background

Modern sewage treament plants aim to reduce the BOD of the incoming material from about 600 mg l^{-1} (five days, 20 °C) to leave an effluent with a BOD of about 20 mg l^{-1}. The processes that they employ to achieve this are similar to the natural purification processes that occur in rivers, but by arranging for them to take place in a man-made environment, the consequent changes in the river community are avoided.

Effluent purification takes place in several stages. Preliminary treatment involves removing large particles on giant metal filters. This is followed by the primary treatment in which smaller particles are allowed to settle out in large tanks. The resulting liquid, called settled sewage, then undergoes biological oxidation in the secondary treatment stage.

In Britain, two types of secondary treatment are used, the activated sludge process and the trickling filter bed. The activated sludge process involves seeding the settled sewage with bacteria and encouraging rapid growth of the bacterial populations by vigorously aerating the sewage in large tanks. Filter beds consist of coarse gravel through which the settled sewage is allowed to percolate. Layers of bacteria and fungi develop on the gravel and these consume the organic material in the sewage. The term 'filter bed' is a misnomer because its main function is to encourage bacterial digestion of organic material, not to act as a physical filter. A flow of air through the pores in the bed satisfies the oxygen demand of the effluent.

The tertiary treatment stage is less common in Britain and is designed to remove inorganic nutrients such as phosphates and nitrates. In this case, a chemical method of treatment is necessary, which is relatively expensive.

Aim

To study the specialised community found in trickling filter beds.

Materials

Plankton net; sample jars; sorting trays; microscope, cavity slides and coverslips; rubber gloves.

Time

1 h in the laboratory plus extra time for visiting the treatment plant.

Method

In Britain sewage treatment works are the responsibility of the regional water authorities. Get the permission of the authority to sample from your local treatment works; often the authority education officer will be able to provide useful information and possibly even arrange a guided tour. When handling material collected from filter beds wear rubber gloves and be sure to observe normal standards of hygiene.

Animals of the filter bed community can be collected by holding a plankton net (mesh size 125 μm) in the drain from the bed or, more directly, by collecting a sample of the bed material. In the latter case about 1 l of bed material is sufficient.

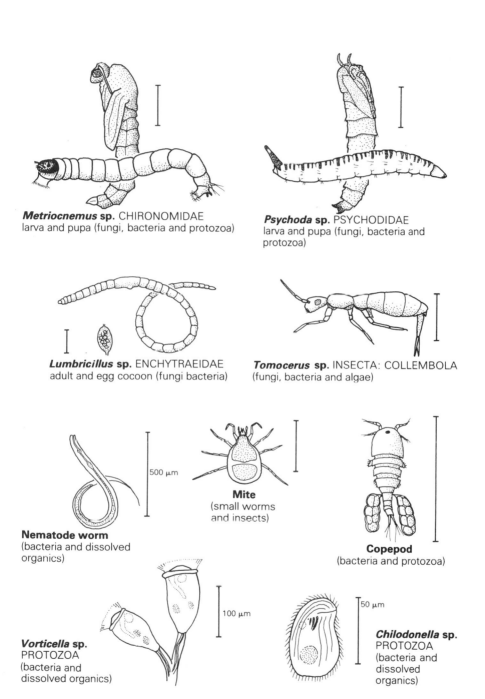

Metriocnemus sp. CHIRONOMIDAE
larva and pupa (fungi, bacteria and protozoa)

Psychoda sp. PSYCHODIDAE
larva and pupa (fungi, bacteria and protozoa)

Lumbricillus sp. ENCHYTRAEIDAE
adult and egg cocoon (fungi bacteria)

Tomocerus sp. INSECTA: COLLEMBOLA
(fungi, bacteria and algae)

500 μm

Nematode worm
(bacteria and dissolved organics)

Mite
(small worms and insects)

Copepod
(bacteria and protozoa)

100 μm

Vorticella sp.
PROTOZOA
(bacteria and dissolved organics)

50 μm

Chilodonella sp.
PROTOZOA
(bacteria and dissolved organics)

Figure 29 Common members of the trickling filter bed community. Unless otherwise specified, the scale bars represent 1 mm.

The very small protozoa are best seen by scraping some of the bacterial film from the surface of the bed material and examining it under about $200 \times$ magnification in a cavity slide. The larger animals can be separated by washing the film on the bed material through a coarse filter paper or a plankton net. Figure 29 will help in the identification of some of the commoner members of the filter bed community and also provides information about their role in sewage treatment.

Questions

(1) Construct a community food web for the filter bed. How important is primary production in this community?

(2) Find out the average volume of effluent passing through the bed and the difference in BOD between the input and the output (the water authority may help you with this). If the breakdown of 1 mg of organic material requires approximately 1.5 mg of oxygen, how much organic material is being processed by the bed each day? What is the final fate of this organic matter?

(3) Efficient functioning of filter beds depends on good drainage and aerobic conditions in the bed. Why are the animals, which graze on the bacterial film on the stones, crucial to filter bed operation?

Appendix 1
Collecting kit

This is a set of equipment which is almost invariably required when freshwater invertebrates are to be sampled, identified and counted. None of the items is expensive or difficult to obtain.

Pond net: square net frames are the most useful since these have a large mouth and a straight edge which can be pressed against the bottom of a stream or pond. The optimum mesh size is usually a compromise between one coarse enough to allow unwanted fine debris to pass through, and one fine enough to retain all the animals of interest. For many purposes a mesh size of 1 mm is adequate for benthic invertebrates, but quantitative work on small species or early instars often requires a mesh of 300 μm or less. If the net is to be used to collect zooplankton, a mesh size of 125 or 150 μm is necessary.

Sample containers: wide mouthed plastic jars are ideal but the ubiquitous polythene bag is cheap and satisfactory. Often freshwater organisms travel better in damp moss than in water. Dissolved oxygen concentrations in warm water can be very low.

Sorting trays: sorting animals from inorganic and organic debris is always a problem. Shallow dishes in which the sample can be spread thinly with a little water are a great help. A white background is useful for spotting many organisms although some of the more transparent species show up best against a dark background. A white dish with one half of the bottom painted black is a useful compromise.

Pipettes: many species are most easily handled if a Pasteur pipette with a mouth at least 4 mm in diameter is available. If the teat is large enough to provide a sudden inflow of water when it is released, it is relatively easy to catch even the fastest moving species.

Specimen trays: the catch from a sample will consist of many individuals of a dozen or more species. These can be kept in separate dishes for identification and counting, but a tray with several compartments or wells is often more convenient. Plastic egg boxes are a cheap source of multi-well trays.

Specimen tubes: these are used for long-term storage of preserved material or for carrying live animals which have been picked out of the net at the water's edge. The familiar 2 × 1 inch glass specimen tubes with plastic push-on closures are quite adequate, although occasionally the closures are apt to pop off, and broken tubes can be dangerous. Polystyrene blood sample tubes (25 cm^3) with screw tops avoid both these problems although they are rather more expensive.

Appendix 2
Measuring the dissolved oxygen content of fresh water

The most convenient way of measuring dissolved oxygen is by using an oxygen meter and electrolytic probe. Unfortunately these are still relatively expensive and often require frequent recalibration in use. The probe has to be carefully maintained if it is to give reliable readings. The paper by Richardson (1981; see Bibliography) gives useful advice to users and purchasers of oxygen meters.

Oxygen determination using the Winkler reaction has the disadvantages that the equipment is not readily portable, the reagents are potentially hazardous and the method is more involved. Nevertheless, it remains the reliable standby when oxygen meters fail to perform, and is the method by which others are judged. The equipment is available in any chemistry laboratory and the reagents are cheap. Gill (1977; see Bibliography) describes a modification of the Winkler method suitable for use in the field.

Equipment

25 cm^3 burette, stand, white tile, filter funnel and 100 cm^3 beaker; 50 cm^3 bulb pipette; 3×5 cm^3 graduated pipettes; pipette safety bulb; 250 cm^3 conical flask.

Reagent solutions

Manganous sulphate (436 g MnSO.4H$_2$O in 1 litre H$_2$O).
Alkaline iodide (500 g NaOH and 135 g NaI in 1 litre H$_2$O).
Sulphuric acid (500 cm^3 conc. H$_2$SO$_4$ in 1 litre H$_2$O–caution!)
Sodium thiosulphate 0.005M (diluted from 0.1M soln – 24.82 g Na$_2$S$_2$O$_3$
.5H$_2$O in 1 litre H$_2$O).
Starch indicator (add 1 g potato starch to 100 cm^3 boiling H$_2$O, stir and
filter through a large fluted paper. Always use fresh starch solution).

Method

The water sample to be analysed should be in a 250 cm^3 glass bottle with a ground-glass stopper and should contain no air bubbles.

(1) Remove the stopper and add 2 cm³ manganous sulphate solution, keeping the tip of the pipette just below the water surface. Add 2 cm³ alkaline iodide solution in the same way.

(2) Replace the stopper so that no air bubbles are trapped in the bottle. Mix the contents thoroughly by inverting several times. A brownish precipitate will form and should be allowed to settle at least halfway down the bottle. At this stage the sample can be stored for a few days, preferably under water.

(3) Remove the stopper and add 4 cm³ sulphuric acid solution. Replace the stopper without trapping air and mix until the precipitate dissolves.

(4) Place 50 cm³ of the sample solution in the conical flask and run in sodium thiosulphate solution from the burette until only a pale straw colour remains.

(5) Add two drops of starch indicator and complete the titration quickly. The end point is reached when the solution turns from blue to colourless. If the end point has not been overshot, the blue colour will return after the solution has stood for about 30 min. Repeat steps 4 and 5 if necessary.

(6) Calculate the oxygen concentration of the sample using the following formula:

$$O_2 \text{ in } mg\,l^{-1} = 8000\,(m/V)v$$

where m is the molarity of the thiosulphate, V is the volume of sample used and v is the volume of thiosulphate titrant.

Strictly, the molarity of the thiosulphate should be checked by standardising against a potassium iodate solution of accurately known strength. A method for doing this is given by Mackereth *et al.* (1978; see Bibliography). In practice this can be left out unless great accuracy is required, providing that the thiosulphate solution has been freshly made up and stored in a refrigerator. For most purposes the correction for the dissolved oxygen displaced by the added reagents can also be ignored.

The strength of the thiosulphate solution, the volume of sample used and the size of the glassware can all be changed to suit individual circumstances. In any event, the formula given in step 6 still applies.

Bibliography

References

Belcher, H. and E. Swale 1976. *A beginner's guide to freshwater algae*. London: HMSO.

Gill, B. F. 1977. A plastic syringe method for measuring dissolved oxygen in the field or a laboratory. *School Sci. Rev.* **58**, 645–8.

Hansell, M. H. and J. J. Aitken 1977. *Experimental animal behaviour*. Glasgow: Blackie.

Haslam, S., C. Sinker, and P. Wolseley 1975. British water plants. *Field Studies* **4**, 243–351.

Macan, T. T. 1965. *A revised key to the British water bugs (Hemiptera–Heteroptera)*. Freshwater Biological Association, publn 16.

Mackereth, F. J. H., J. Heron and J. F. Talling 1978. *Water analysis*. Freshwater Biological Association, publn 36.

Moore, J. 1984. Parasites that change the behaviour of their host. *Scient. Am.* **250**, 82–9.

Richardson, J. 1981. Oxygen meters: some practical considerations. *J. Biol. Education* **15**, 107–16.

Scourfield, D. J. and J. P. Harding 1958. *A key to the British species of freshwater Cladocera*. Freshwater Biological Association, publn 5.

Further reading

Clegg, J. 1974. *Freshwater life*. London: Frederick Warne.

Englehardt, W. and H. Merxmuller 1973. *Pond life*. London: Burke.

Hynes, H. B. N. 1960. *The biology of polluted waters*. Liverpool: Liverpool University Press.

Hynes, H. B. N. 1970. *The ecology of running waters*. Liverpool: Liverpool University Press.

King, T. J. 1980. *Ecology*. Walton-on-Thames: Nelson.

Leadley Brown, A. 1971. *Ecology of fresh water*. London: Heinemann Educational.

Macan, T. T. 1973. *Ponds and lakes*. London: George Allen & Unwin.

Macan, T. T. 1974. *Freshwater ecology*. London: Longman.

Macan, T. T. and E. B. Worthington 1974. *Life in lakes and rivers*. London: Collins.

Maitland, P. S. 1978. *Biology of fresh waters*. Glasgow: Blackie.

Mason, C. F. 1981. *Biology of freshwater pollution*. London: Longman.

Mellanby, H. 1975. *Animal life in fresh water*. London: Chapman and Hall.

Moss, B. 1980. *Ecology of fresh waters*. Oxford: Blackwell Scientific.

Sterry, P. 1983. *Pond watching*. London: Severn House.

Thompson, G., J. Coldrey and G. Bernard 1984. *The Pond*. London: Collins.

Townsend, C. R. 1980. *The ecology of streams and rivers*. London: Edward Arnold.

Keys

Edington, J. M. and A. G. Hildrew 1981. *Caseless caddis larvae of the British Isles.* Freshwater Biological Association, publn 43.

Elliot, J. M. and K. H. Mann 1979. A key to the British freshwater leeches. Freshwater Biological Association, publn 40.

Gledhill, T., D. W. Sutcliffe and W. D. Williams 1976. *A key to the British freshwater Crustacea: Malacostraca.* Freshwater Biological Association, publn 32.

Hammond, D. 1983. *Dragonflies of Great Britain and Ireland.* Colchester: Harley Books.

Hickin, N. E. 1967. *Caddis larvae: larvae of the British Trichoptera.* London: Hutchinson.

Holland, D. G. 1972. *A key to the larval, pupal and adults of the British species of Elminthidae.* Freshwater Biological Association, publn 26.

Hynes, H. B. N. 1977. *A revised key to the adults and nymphs of British stoneflies (Plecoptera).* Freshwater Biological Association, publn 17.

Macan, T. T. 1959. *A guide to freshwater invertebrate animals.* London: Longman.

Macan, T. T. 1977. *A key to British fresh- and brackish-water gastropods.* Freshwater Biological Association, publn 13.

Macan, T. T. 1979. A key to the nymphs of British Ephemeroptera, Freshwater Biological Association, publn 20.

Maitland, P. S. 1972. *A key to British freshwater fishes.* Freshwater Biological Association, publn 27.

Quigley, M. 1977. *Invertebrates of streams and rivers: a key to identification.* London: Edward Arnold.

Redfern, M. 1975. Revised field key to the invertebrate fauna of stony hill streams. *Field Studies* **4** (2), 105–15.

Reynoldson, T. B. 1978. *A key to the British species of freshwater triclads.* Freshwater Biological Association, publn 23.

Freshwater Biological Association publications are obtainable from: The Librarian, The Ferry House, Ambleside, Cumbria LA22 02P.

Index

Italic numbers refer to text figures, and bold numbers refer to text sections.